INTERNATIONAL CONFERENCE ON NUCLEAR SECURITY: SHAPING THE FUTURE

The following States are Members of the International Atomic Energy Agency:

AFGHANISTAN
ALBANIA
ALGERIA
ANGOLA
ANTIGUA AND BARBUDA
ARGENTINA
ARMENIA
AUSTRALIA
AUSTRIA
AZERBAIJAN
BAHAMAS, THE
BAHRAIN
BANGLADESH
BARBADOS
BELARUS
BELGIUM
BELIZE
BENIN
BOLIVIA, PLURINATIONAL
    STATE OF
BOSNIA AND HERZEGOVINA
BOTSWANA
BRAZIL
BRUNEI DARUSSALAM
BULGARIA
BURKINA FASO
BURUNDI
CABO VERDE
CAMBODIA
CAMEROON
CANADA
CENTRAL AFRICAN
    REPUBLIC
CHAD
CHILE
CHINA
COLOMBIA
COMOROS
CONGO
COOK ISLANDS
COSTA RICA
CÔTE D'IVOIRE
CROATIA
CUBA
CYPRUS
CZECH REPUBLIC
DEMOCRATIC REPUBLIC
    OF THE CONGO
DENMARK
DJIBOUTI
DOMINICA
DOMINICAN REPUBLIC
ECUADOR
EGYPT
EL SALVADOR
ERITREA
ESTONIA
ESWATINI
ETHIOPIA
FIJI
FINLAND
FRANCE
GABON
GAMBIA, THE

GEORGIA
GERMANY
GHANA
GREECE
GRENADA
GUATEMALA
GUINEA
GUYANA
HAITI
HOLY SEE
HONDURAS
HUNGARY
ICELAND
INDIA
INDONESIA
IRAN, ISLAMIC REPUBLIC OF
IRAQ
IRELAND
ISRAEL
ITALY
JAMAICA
JAPAN
JORDAN
KAZAKHSTAN
KENYA
KOREA, REPUBLIC OF
KUWAIT
KYRGYZSTAN
LAO PEOPLE'S DEMOCRATIC
    REPUBLIC
LATVIA
LEBANON
LESOTHO
LIBERIA
LIBYA
LIECHTENSTEIN
LITHUANIA
LUXEMBOURG
MADAGASCAR
MALAWI
MALAYSIA
MALI
MALTA
MARSHALL ISLANDS
MAURITANIA
MAURITIUS
MEXICO
MONACO
MONGOLIA
MONTENEGRO
MOROCCO
MOZAMBIQUE
MYANMAR
NAMIBIA
NEPAL
NETHERLANDS,
    KINGDOM OF THE
NEW ZEALAND
NICARAGUA
NIGER
NIGERIA
NORTH MACEDONIA
NORWAY
OMAN

PAKISTAN
PALAU
PANAMA
PAPUA NEW GUINEA
PARAGUAY
PERU
PHILIPPINES
POLAND
PORTUGAL
QATAR
REPUBLIC OF MOLDOVA
ROMANIA
RUSSIAN FEDERATION
RWANDA
SAINT KITTS AND NEVIS
SAINT LUCIA
SAINT VINCENT AND
    THE GRENADINES
SAMOA
SAN MARINO
SAUDI ARABIA
SENEGAL
SERBIA
SEYCHELLES
SIERRA LEONE
SINGAPORE
SLOVAKIA
SLOVENIA
SOMALIA
SOUTH AFRICA
SPAIN
SRI LANKA
SUDAN
SWEDEN
SWITZERLAND
SYRIAN ARAB REPUBLIC
TAJIKISTAN
THAILAND
TOGO
TONGA
TRINIDAD AND TOBAGO
TUNISIA
TÜRKİYE
TURKMENISTAN
UGANDA
UKRAINE
UNITED ARAB EMIRATES
UNITED KINGDOM OF
    GREAT BRITAIN AND
    NORTHERN IRELAND
UNITED REPUBLIC OF TANZANIA
UNITED STATES OF AMERICA
URUGUAY
UZBEKISTAN
VANUATU
VENEZUELA, BOLIVARIAN
    REPUBLIC OF
VIET NAM
YEMEN
ZAMBIA
ZIMBABWE

The Agency's Statute was approved on 23 October 1956 by the Conference on the Statute of the IAEA held at United Nations Headquarters, New York; it entered into force on 29 July 1957. The Headquarters of the Agency are situated in Vienna. Its principal objective is "to accelerate and enlarge the contribution of atomic energy to peace, health and prosperity throughout the world".

PROCEEDINGS SERIES

INTERNATIONAL CONFERENCE ON NUCLEAR SECURITY: SHAPING THE FUTURE

PROCEEDINGS OF AN INTERNATIONAL CONFERENCE ORGANIZED BY THE
INTERNATIONAL ATOMIC ENERGY AGENCY
AND HELD IN VIENNA, 20–24 MAY 2024

INTERNATIONAL ATOMIC ENERGY AGENCY
VIENNA, 2025

IAEA Library Cataloguing in Publication Data

Names: International Atomic Energy Agency.
Title: International conference on nuclear security : shaping the future / International Atomic Energy Agency.
Description: Vienna : International Atomic Energy Agency, 2025. | Series: Proceedings series (International Atomic Energy Agency), ISSN 0074-1884 | Includes bibliographical references.
Identifiers: IAEAL 25-01800 | ISBN 978-92-0-124525-0 (paperback : alk. paper) | ISBN 978-92-0-124625-7 (pdf)
Subjects: LCSH: Nuclear industry — Security measures — Congresses. | Nuclear industry — International cooperation. | Nuclear nonproliferation.
Classification: UDC 341.67 | STI/PUB/2128

FOREWORD

The International Conference on Nuclear Security: Shaping the Future (ICONS 2024) was held at IAEA Headquarters in Vienna from 20 to 24 May 2024. It was the fourth quadrennial ministerial conference on nuclear security convened by the IAEA, following the conferences held in July 2013, December 2016 and February 2020. A high level of attendance was sustained throughout the week, with participants representing a broad range of entities that support or are actively engaged in nuclear security activities worldwide, including regulatory bodies, competent authorities, national security and crisis management agencies, law enforcement and border control agencies, industry, and intergovernmental and non-governmental organizations.

The conference attracted more than 2000 registered participants, including 11 ministers, 24 vice-ministers and 14 other high ranking officials, underscoring its role as the largest and foremost global forum for nuclear security practitioners. This turnout also demonstrated the widespread recognition within the international community that nuclear security is a topic that needs continued attention, discussion and international dialogue. Conference panellists and participants consistently emphasized the value of periodic international workshops, meetings and forums, such as ICONS, to share best practices and foster sustained and meaningful international cooperation.

The conference's core theme of 'Shaping the Future' inspired many of the activities and discussions included in the programme. Preference was given to papers and side events that addressed emerging threats and explored future challenges. This general orientation towards the future allowed participants to share experiences, achievements and lessons identified since the International Conference on Nuclear Security: Sustaining and Strengthening Efforts (ICONS 2020), while also shifting the international community's focus to potential challenges ahead. ICONS 2024 thus served as a vehicle through which ministers, policy makers, senior officials and nuclear security experts could continue developing shared approaches to existing threats while also formulating cooperative strategies to prepare for the challenges of the future.

These proceedings include the Co-Presidents' Report of the conference, the full text of the Statement by the Co-Presidents, statements from the opening and closing sessions, and lists of all papers and flash presentations (short oral presentations) presented as part of the scientific and technical programme. The conference programme, the list of conference participants and the photographs featured in the photography competition entitled 'Nuclear Security Through the Lens' are available on-line as supplementary files.

The IAEA is grateful for the cooperation and support of the numerous organizations and individuals involved in the planning and implementation of this conference. The IAEA officers responsible for this publication were S. Mroz, B. Denehy and C. Deura of the Division of Nuclear Security.

CONTENTS

1. INTRODUCTION 1
   1.1. BACKGROUND 1
   1.2. OVERVIEW OF THE CONFERENCE 2
      1.2.1. Ministerial segment 2
      1.2.2. Scientific and technical segment 3

2. OPENING ADDRESSES 4
   2.1. INTERNATIONAL ATOMIC ENERGY AGENCY 4
   2.2. CONFERENCE CO-PRESIDENT, KAZAKHSTAN 7
   2.3. CONFERENCE CO-PRESIDENT, AUSTRALIA 9
   2.4. OUTGOING CONFERENCE PRESIDENT, ROMANIA 12
   2.5. OUTGOING CONFERENCE PRESIDENT, PANAMA 15

3. CLOSING ADDRESSES 16
   3.1. CONFERENCE CO-PRESIDENT, AUSTRALIA 16
   3.2. CONFERENCE CO-PRESIDENT, KAZAKHSTAN 18
   3.3. INTERNATIONAL ATOMIC ENERGY AGENCY 20

4. CO-PRESIDENTS' REPORT 21
   4.1. INTRODUCTION 21
   4.2. MINISTERIAL SESSION 21
      4.2.1. Opening plenary session 21
      4.2.2. National statements 22
      4.2.3. Plenary panel: Securing sustainable progress — The important role of nuclear security in advancing the sustainable development goals 22
      4.2.4. Ministerial segment event: Beyond borders — A collaborative discourse on the future of nuclear security 22
      4.2.5. Interactive ministerial session 23
   4.3. SCIENTIFIC AND TECHNICAL SEGMENT 23
      4.3.1. Opening plenary session 23
   4.4. POLICY, LAW AND REGULATIONS FOR NUCLEAR SECURITY 23
      4.4.1. Plenary panel: Policy, law and regulations in an evolving nuclear security landscape 23
      4.4.2. Technical sessions 24
   4.5. TECHNOLOGY AND INFRASTRUCTURE FOR NUCLEAR SECURITY PREVENTION, DETECTION AND RESPONSE 26
      4.5.1. Plenary panel: Managing the threats and benefits of emerging technologies . 27
      4.5.2. Technical sessions 27
   4.6. CAPACITY BUILDING FOR NUCLEAR SECURITY 32
      4.6.1. Plenary panel: Shaping the future — Retaining and developing the nuclear security workforce of tomorrow 32
      4.6.2. Technical sessions 33
   4.7. CROSS-CUTTING NUCLEAR SECURITY TOPICS 35
      4.7.1. Plenary panel: Looking forward — The evolving role of the IAEA 35
      4.7.2. Technical sessions 36
   4.8. CLOSING PLENARY SESSION 38

5. STATEMENT BY THE CO-PRESIDENTS 40

6. STATEMENT BY THE NUCLEAR SECURITY DELEGATION FOR THE FUTURE 43

7. CONTENTS OF THE SUPPLEMENTARY FILES 46

ANNEX I.   CONFERENCE STATISTICAL DATA 47

ANNEX II.   LIST OF TECHNICAL AND SCIENTIFIC PAPERS 51

ANNEX III.   LIST OF FLASH PRESENTATIONS 95

CONFERENCE ORGANIZERS 107

1. INTRODUCTION

1.1. BACKGROUND

Since the early 1970s, the IAEA has been providing assistance to its Member States, upon request, in support of their national efforts to establish and strengthen nuclear security. In March 2002, the IAEA's Board of Governors approved the first comprehensive action plan to protect against nuclear terrorism. Further IAEA Nuclear Security Plans were approved in 2005, 2009, 2013, 2017 and 2021. Member States have consistently recognized the central role of the IAEA in strengthening the nuclear security framework globally and in coordinating international cooperation in nuclear security. The objectives of the IAEA Nuclear Security Programme, as set out in the current Nuclear Security Plan 2022–2025 (GOV/2021/34)[1], are as follows:

— To assist States in establishing, maintaining and sustaining national nuclear security regimes for nuclear and other radioactive materials, including during transport, and associated facilities used for peaceful purposes;
— To contribute to global efforts to achieve effective nuclear security, including by establishing comprehensive nuclear security guidance and, upon request, promoting its use through peer reviews, advisory services and capacity building, including education and training;
— To promote adherence to relevant international legal instruments and commitment to the Code of Conduct on the Safety and Security of Radioactive Sources[2] and its supplementary guidance to enhance nuclear security globally;
— To play the central role of facilitating and enhancing international cooperation and increasing visibility and awareness through communication on nuclear security, in response to resolutions and decisions of the General Conference and Board of Governors, and considering Ministerial Declarations.

The International Conference on Nuclear Security: Shaping the Future (ICONS 2024), held at IAEA Headquarters in Vienna from 20 to 24 May 2024, was the fourth quadrennial ministerial conference on nuclear security convened by the IAEA. It provided a global forum for ministers, policy makers, senior officials and nuclear security experts to discuss the future of nuclear security worldwide, while providing opportunities to exchange information, share best practices and foster international cooperation.

A wide range of entities that support or are actively engaged in nuclear security activities worldwide were represented at the conference from regulatory bodies, competent authorities, and national security and crisis management agencies to law enforcement and border control agencies, industry, and intergovernmental and non-governmental organizations. Outputs from discussions at the conference will be used to inform the preparation of the next IAEA Nuclear Security Plan, which will cover the period 2026–2029.

ICONS 2024 sought specifically to undertake the following:

— Raise awareness on a wide range of nuclear security topics to maintain and further strengthen national nuclear security regimes for nuclear and other radioactive material and associated facilities used for peaceful purposes, as well as international cooperation in strengthening nuclear security globally;
— Review nuclear security experiences and achievements, as well as current approaches and trends, and highlight areas that may need more focused attention, including technological dimensions;
— Promote IAEA nuclear security guidance, and the experience of States in its application, including through peer reviews, advisory services and capacity building;
— Promote the sharing of information and good practices in nuclear security while protecting sensitive information;

---

[1] Nuclear Security Plan 2022–2025, Board of Governors document GOV/2021/34, IAEA, Vienna (2021), https://www.iaea.org/sites/default/files/gc/gc65-24.pdf
[2] INTERNATIONAL ATOMIC ENERGY AGENCY, Code of Conduct on the Safety and Security of Radioactive Sources, IAEA, Vienna (2004)

— Reaffirm and support the central role of the IAEA in strengthening nuclear security globally and in leading the coordination of international activities in the field of nuclear security, while avoiding duplication and overlap;
— Highlight and promote IAEA efforts to promote adherence to relevant international legally binding instruments and commitment to legally non-binding instruments;
— Discuss further enhancements to IAEA nuclear security activities and their sustainability.

More than 2000 participants from 142 Member States and 16 invited organizations registered to participate at ICONS 2024, with 11 ministers, 24 vice-ministers and 14 other high ranking officials participating in the ministerial segment of the conference.

The Secretariat made a concerted effort throughout all phases of planning the conference to ensure that participants represented a range of backgrounds and levels of expertise, from senior policy makers, industry leaders and established nuclear security specialists to emerging experts and future leaders in the field. A total of 142 Member States and 16 invited organizations participated in the conference, and 34% of registered participants were women. Representatives from 89 Member States delivered presentations during the scientific and technical segment of the conference. To ensure diversity, not only in terms of the participants but also in terms of the ideas and perspectives discussed during panels and sessions, the scientific and technical sessions were moderated by session chairs representing 41 Member States and invited organizations.

The Co-Presidents' Report, included as Section 4 of these proceedings, highlights the key issues and main conclusions from the conference.

1.2. OVERVIEW OF THE CONFERENCE

The conference was divided into two parts. The first part of the conference consisted of two days focused on the ministerial segment that aimed at providing an opportunity for ministers to engage in discussions on nuclear security and to deliver messages on achievements and priorities. The second part, from days two through five, featured the scientific and technical segment of the conference, comprising high level plenary policy discussions on the overall themes of the conference and parallel technical sessions on specialized scientific, technical, legal and regulatory issues concerning nuclear security.[3]

1.2.1. Ministerial segment

The Assistant Minister for Foreign Affairs of Australia, HE Mr Tim Watts, MP; and the Vice Minister of Energy of Kazakhstan, HE Mr Sungat Yessimkhanov, assumed the role of Co-Presidents of the conference. The Ambassadors and Permanent Representatives of Australia and Kazakhstan, HE Mr Ian Biggs and HE Mr Mukhtar Tileuberdi respectively, co-chaired the informal open ended consultations among Member States, leading to the Statement by the Co-Presidents on nuclear security.

The conference opened with addresses by the IAEA Director General, Mr Rafael Mariano Grossi, and by the Co-Presidents of the conference. HE Ms Ana Cristina Tinca, Deputy Minister for Foreign Affairs and State Secretary for Strategic Affairs of Romania; and HE Mr Dario Chiru Ochoa, Ambassador and Permanent Representative of Panama, both provided remarks in their capacities as outgoing Co-Presidents of ICONS 2020, reflecting on developments in nuclear security since that conference.

During the opening session, the Co-Presidents presented the Statement by the Co-Presidents in lieu of a Ministerial Declaration on Nuclear Security. While many Member States agreed on the tenets of a draft Ministerial Declaration that had been negotiated in the months preceding the conference, consensus was not achieved. The Statement by the Co-Presidents thus sought to reflect the outcome of these extensive negotiations and open ended working groups, highlighting the crucial role that nuclear security plays in

---

[3] The full programme of the conference is available https://www.iaea.org/events/icons2024 in the supplementary files to this publication, available on the IAEA website.

achieving global peace and security and reaffirming the central role of the IAEA in coordinating international nuclear security activities. The Co-Presidents requested that Member States indicate their alignment with the Statement by the Co-Presidents during their national remarks.

A total of 99 national statements were delivered by ministers and other heads of delegation, along with 3 joint statements, 2 statements from representatives of international organizations, and 1 State exercising its right of reply. All of the speakers acknowledged the importance of national commitments to strengthen nuclear security globally, and the need for international cooperation and assistance to complement and support national actions. Many also expressed appreciation for the central role of the IAEA in coordinating such international efforts and for providing assistance when requested.

In addition to national and organizational statements, the ministerial segment of the conference featured opportunities for participants to engage in discussions relating to specific topics concerning nuclear security. A plenary panel discussion focused on the importance of nuclear security in advancing the peaceful uses of nuclear and other radioactive material in the pursuit of the United Nations Sustainable Development Goals. An evening panel discussion examined the need for, and the challenges associated with, international cooperation, particularly in relation to the development and adoption of policies and legislation. A third event offered an interactive, scenario based policy discussion, in which participants were led through a scenario and a moderated discussion about decision points. Each of these sessions enabled ministers and other heads of delegations to participate in discussions, meet fellow leaders and deepen their understanding of nuclear security topics.

1.2.2. Scientific and technical segment

The scientific and technical segment of the conference opened on the second day of the conference and ran in parallel to the second day of the ministerial segment of the conference. During the opening plenary session for the scientific and technical segment, remarks were delivered by the IAEA Deputy Director General and Head of the Department of Nuclear Safety and Security, Ms Lydie Evrard; the Director of the IAEA Division of Nuclear Security, Ms Elena Buglova; and the Scientific Secretary of ICONS 2024, Ms Sara Mroz.

During this session, the IAEA Secretariat also introduced the Nuclear Security Delegation for the Future. This cohort of young nuclear security professionals and students interested in pursuing careers in nuclear security was launched during the preparations for ICONS 2024. The group includes 24 participants from 19 States. Members of the group took an active role in the conference and were provided with various professional development opportunities.

The scientific and technical segment was organized around four main themes encompassing various aspects of nuclear security. These four themes were as follows:

— Policy, law and regulations for nuclear security;
— Technology and infrastructure for nuclear security prevention, detection and response;
— Capacity building for nuclear security;
— Cross-cutting nuclear security topics.

Each day of the scientific and technical segment featured an opening plenary session dedicated to one of the four thematic areas listed above. The four plenary sessions featured moderated discussions with experts from around the world to highlight important challenges and ideas for the future of nuclear security.

Technical sessions consisted of presentations and discussions on detailed technical topics, organized by thematic area. In total, the conference programme comprised 52 technical sessions, along with a flash presentation event that featured 60 presentations. The scientific and technical segment was also complemented by 45 side events and 50 exhibits hosted by Member States, organizations and the IAEA.

2. OPENING ADDRESSES

2.1. INTERNATIONAL ATOMIC ENERGY AGENCY

Mr Rafael Mariano Grossi

Director General
International Atomic Energy Agency

When we met the last time, at ICONS 2020, many of us could not have imagined the momentous change we would experience between then and today, change that would affect billions of people, international peace and security, and nuclear security. A global pandemic was in the making and a war — in Ukraine — for the first time would soon be fought among the facilities of one of Europe's biggest nuclear power programmes.

Meanwhile, profound technological advances have been made. Assessing their impact on nuclear security is a crucial task. Artificial intelligence (AI) and unmanned vehicles pose both a threat to nuclear security and offer new tools with which to enhance it. In the nuclear field itself, small modular reactors promise new opportunities for applications, such as desalination and power brought to remote communities via barge, but also require us to consider new security elements.

The use of nuclear science and technology, often facilitated by the IAEA, has come on in leaps and bounds. Climate change and the drive for energy security are fuelling a desire for nuclear power. At this past Conference of the Parties to the UN Framework Convention on Climate Change, COP28, world leaders — those whose States use nuclear power and those whose do not — for the first time in nearly 30 years of COP meetings agreed nuclear power must be part of the transition to net zero. More than 20 countries have signed a pledge towards tripling nuclear power capacity, and at the IAEA Nuclear Energy Summit in March, heads of State agreed on the urgent need for conducive financial conditions.

Nuclear security is relevant throughout all the steps of the nuclear fuel cycle and is part of the social contract that underpins the existence and growth of nuclear power. Nuclear power programmes require national nuclear security threat assessments and 'security by design'. Nurturing relevant research and a strong security culture are key, not only in countries with NPPs.

The use of life-saving and life-affirming applications of nuclear science and technology is growing, from cancer patients gaining access to radiotherapy to farmers benefiting from new crop varieties developed with the help of irradiation. IAEA initiatives, such as Rays of Hope: Cancer care for all; NUTEC Plastics; Zoonotic Disease Integrated Action (ZODIAC); and Atoms4Food are key vehicles facilitating wider access.

All these opportunities to use nuclear and radioactive material depend on a strong and adaptive global nuclear security regime. For countries new to using nuclear and radioactive material, this means building up legal infrastructure, practices and culture that bolster nuclear security. Nationally and across borders, collaboration and laser-focused vigilance are key to preventing groups with malicious intent from using nuclear and other radioactive material to cause panic and harm.

The threats to nuclear and other radioactive material and associated facilities are real and varied. The international nuclear security threat landscape keeps evolving. Today, anyone can type a few words into a computer and generative AI can create images of nuclear Armageddon, meaning it is now possible to spread panic about radiation fallout without a nuclear device. Risk scenarios include theft of nuclear and other radioactive material for use in improvised devices and sabotage at nuclear installations or during transport of nuclear and radioactive material. The risk of cyber-attacks requires the implementation of computer security programmes by those who use nuclear power and those who don't. Risks come from outsiders and from those within the fold who are disgruntled or have been corrupted.

Nuclear security is the national responsibility of individual States, but it also benefits enormously from close collaboration and the enabling role of the IAEA. ICONS, which started in 2013, has been the place for ministers, policy makers, senior officials and experts to gather to assesses current priorities, prepare for new challenges and engage in scenario based policy discussions. ICONS 2024, presided over by the Co-Presidents, HE Tim Watts, Assistant Minister for Foreign Affairs of Australia, and HE Sungat Yessimkhanov, Vice Minister of Energy of Kazakhstan, covers the themes of policy, law and regulation; technology and infrastructure for prevention, detection and response; capacity building; and cross-cutting areas, such as the interface between nuclear security and nuclear safety. ICONS is the most important high level international meeting on nuclear security. At this time of heightened tensions, it is imperative that there remains a unity of purpose and that nuclear security does not become a political football.

This year marks the 10 year anniversary of the IAEA's Division of Nuclear Security. The IAEA is at the forefront of adapting nuclear security to new challenges, including war. The seven indispensable pillars for ensuring nuclear safety and security have broad international support. They have brought crucial clarity at a time of war and are testament to the adaptiveness of the IAEA and the security regime.

Those seven pillars are backed up by an enormous ongoing effort by the IAEA to support Ukraine, including through the continuous presence of IAEA experts at all of Ukraine's nuclear power plants, including Zaporizhzhya NPP on the front lines of the war. When there were allegations of nuclear security breaches, the IAEA was there to investigate with impartiality and science. We set the facts straight that no nuclear material had been diverted, cutting through the fog of war, and diffusing a tense situation.

Not all our efforts require quite as much courage as our experts have shown in Ukraine, nor do they make international headlines. But every day, the IAEA — the Secretariat and the Member States — work together fastidiously to underpin nuclear security, never resting, always learning.

Radioactive sources are extensively used in many domains, including medicine, industry, agriculture and research. An incident in one State can have far reaching consequences for others, so security for one is security for all. That means supporting States with no, or less developed nuclear security infrastructure makes everyone safer. That support, which often comes via the IAEA, includes making lawmakers aware of their responsibilities.

Nuclear security requires the implementation of appropriate and robust legislative regulatory frameworks. In 2022, the first Conference of the Parties to the Amendment to the Convention on the Physical Protection of Nuclear Material (A/CPPNM) was held under the auspices of the IAEA. Reflecting the global importance of the legal framework and of nuclear security, parties managed to agree an outcome document and for the IAEA to convene a subsequent conference. Since 2020, 14 new parties have joined the A/CPPNM bringing the total to 136. Five new parties joined the CPPNM, bringing that total to 164. In addition to the A/CPPNM, political commitment to legally non-binding instruments, like the Code of Conduct on the Safety and Security of Radioactive Sources and its supplementary guidance, is a strong indication of radiation safety and nuclear security culture.

But legal frameworks are just the beginning. They must be implemented. The IAEA plays a central role in assisting its Members States so they are able to do that. Last year we inaugurated the most visible symbol of our collaboration: the Nuclear Security Training and Demonstration Centre (NSTDC). This first of its kind space, made possible by 15 donors, is a cornerstone for capacity building amid the growing need for sophisticated hands-on nuclear security training using advanced, specialized equipment. The NSTDC is part of a wide range of services offered by the IAEA, including peer reviews, such as the International Physical Protection Advisory Service (IPPAS), of which there have now been more than 100, and Advisory Missions on Regulatory Infrastructure for Radiation Safety and Nuclear Security (RISS), a service we launched in 2022. Our Incident and Trafficking Database (ITDB) now has 145 members and has enabled the reporting of more than 600 incidents in which nuclear or radioactive material went out of regulatory control. Almost 8000 people have benefited from our training in nuclear security, and we continue to work very hard to remove barriers that prevent talent from entering the field. In March 2021, we launched the Women in Nuclear Security Initiative (WINSI) to support the achievement of gender equality in nuclear security. Meanwhile, the IAEA Marie Skłodowska-Curie Fellowship Programme financially supports women pursuing a master's degree in nuclear subjects and offers them internships, while our Lise Meitner Programme offers women in the early and middle part of their career enriching opportunities within the field.

As the use of nuclear and other radioactive material around the world increases, more and more States are needing to increase their level of nuclear security. Nuclear security is as important as nuclear safety – we must put it on equal footing in terms of reliability of funding and the robustness of implementation.

At ICONS 2024 we are — as the name of the conference indicates — 'shaping the future', not only of nuclear security but of the world our children will inherit. That is because nuclear security is about more than preventing nuclear terrorism. It is an enabler to providing, through nuclear science and technology, the clean energy; cutting-edge medicine; nutritious food and hope for a better tomorrow.

HE Mr Sungat Yessimkhanov

Vice Minister of Energy of Kazakhstan

Mr Co-President, Mr Director General,

Distinguished Excellencies and colleagues,

It is my privilege and honour to welcome you all today as we open the International Conference on Nuclear Security, symbolically entitled 'Shaping the Future'. This is indeed the forum of ministers, policy makers and nuclear security experts from around the world, who are committed to shaping our better future in such an important field as nuclear security.

In this regard, I am pleased to co-chair this important event with Mr Tim Watts, Assistant Minister for Foreign Affairs of Australia, and to thank both the Permanent Missions of Australia and Kazakhstan for their joint efforts in preparing the ministerial segment of the conference.

My sincere gratitude also goes to the Co-Chairs of the Programme Committee for their skilful leadership in preparing the scientific and technical programme of ICONS 2024. I also appreciate the personal efforts of Director General Rafael Mariano Grossi and his professional team.

Dear ladies and gentlemen,

This ICONS comes at a critical time for international nuclear security. The international community today faces many different challenges. At the same time, products of artificial intelligence and advanced computing technologies offer new opportunities to strengthen national nuclear security regimes. As new technologies continue to evolve, we must adapt our nuclear security measures to mitigate potential risks and reap the benefits of these technologies. In this regard, the establishment of robust national nuclear security regimes and the universalization of core legal instruments will ensure a guaranteed nuclear energy chain for humanity. Kazakhstan is a party to the main legal instruments in nuclear security, and we call upon Member States to adhere to them. We also commend the IAEA's role in preventing and responding to acts and threats of nuclear terrorism, in developing nuclear security recommendations and guidelines, and in providing advisory services and capacity building to Member States.

Dear friends,

Since its independence, Kazakhstan has pursued an effective national model to achieve a nuclear-weapon-free world. By voluntarily renouncing nuclear weapons in 1991, the concept of disarmament, non-proliferation and peaceful use of nuclear energy has become an essential part of Kazakhstan's national identity.

Ensuring the nuclear security of the former Semipalatinsk nuclear test site was the first major challenge for newly independent Kazakhstan. Together with our international partners, we have sealed all nuclear test tunnels and shafts, removed weapons-usable material, and secured the vast area of the test site against unauthorized access and illegal transfers. All other nuclear facilities used for peaceful purposes have been properly secured and protected from potential threats.

Kazakhstan fully supports the UN's urgent call for concrete action to protect our environment for future generations. As the world's leading exporter of uranium, providing 43% of the global supply, Kazakhstan plays a crucial role in carbon free power generation on a global scale.

Kazakhstan made a bold leap forward by adopting the Strategy of the Republic of Kazakhstan on Achieving Carbon Neutrality by 2060. Under the Strategy, we pay special attention to developing nuclear and hydrogen energy, renewable energy, and converting coal fuelled facilities to gas.

As Kazakhstan embarks on its first nuclear power programme, we are committed to strictly following IAEA guidelines and recommendations to build a robust nuclear safety and security environment around the future nuclear power plant. Kazakhstan has contributed to the non-proliferation regime and sustainable development of nuclear energy by hosting the unique IAEA Low Enriched Uranium Bank. Moreover, we continue to implement research reactor conversion projects aimed at converting high enriched uranium fuel to low enriched uranium fuel, thereby helping to reduce the risk of nuclear proliferation. Just last year, we successfully completed the conversion of another research reactor at the National Nuclear Center of Kazakhstan, which is now fully operating on low enriched fuel.

Excellencies,

The ICONS 2024 platform, under the auspices of the IAEA, provides a unique opportunity to reaffirm, at the ministerial level, our commitments to promote the peaceful uses of nuclear energy and to meet all necessary international obligations in the field of nuclear security. Through our collective efforts, we can ensure that nuclear technologies are used safely for the benefit of humankind and shape a better future for generations to come. I thank you!

HE Mr Tim Watts MP

Assistant Minister for Foreign Affairs of Australia

Director General, Fellow Co-President, Ministers, Distinguished Colleagues,

It is my pleasure to welcome you here to the 2024 International Conference on Nuclear Security – ICONS.

The past four years have been a time of momentous change for nuclear security.

Far from abating, global tensions have evolved into new, more complex conflict environments, with newer weapons, and diverse, sophisticated actors.

In parallel, there has been a remarkable growth in the use of nuclear technology, in recognition of its benefits for humankind.

In this context the risk that malign actors will access and misuse nuclear material has increased.

A strong and sustainable nuclear security system has never been more imperative.

If there was ever a time for strong ministerial commitment to strengthening nuclear security, that time is now.

Our attendance at ICONS — a key event for our global nuclear security community — signals our shared commitment to strengthening nuclear security.

It provides an opportunity for us to progress aligned commitments and priorities, and to work closely on our respective national nuclear security regimes.

I would like to acknowledge the sustained efforts of Member States to arrive at an ambitious Ministerial Declaration.

I would like to thank Kazakhstan for co-chairing the negotiation process with Australia and welcome the active engagement and flexibility of Member States in these negotiations.

It is disappointing that despite being very close to reaching consensus, we were ultimately unable to achieve agreement.

The Co-Presidents' statement reflects a compromise agreed to by the majority of us here today and captures the political will and commitment of Member States to further strengthen nuclear security into the future.

It is my hope, as Co-President, that alignment with this Co-Presidents' statement will guide and further strengthen efforts of Member States, as well as the work of the IAEA, for enhancing nuclear security into the future.

Distinguished Delegates,

Nuclear science and technology can make a positive and considerable difference to people's lives.

As a world leading producer of nuclear medicines to diagnose and treat cancer and other diseases, Australia recognizes the important role of nuclear security in facilitating the peaceful uses of the atom, in building public confidence in the use of nuclear technology and in ensuring the protection of our communities.

Australia recognizes the unique expertise and role the IAEA plays.

The IAEA's central role of fostering international cooperation in nuclear security is crucial.

We particularly welcome the increasing number of capacity building activities offered by the IAEA.

Fair access to knowledge and resources is a crucial aspect of future planning, fortifying against nuclear security threats, and seizing opportunities.

This principle is evident in Australia's collaborative efforts with our regional partners on nuclear security. In April this year, Australia hosted an IAEA nuclear security workshop that brought together 20 Indo-Pacific Island nations.

The workshop provided a forum to share experiences and encourage the uptake of the IAEA's Integrated Nuclear Security Sustainability Plan mechanism and IAEA membership among Indo-Pacific countries.

One of the key outcomes of the workshop was the realization for many of the Island States that nuclear security is relevant for their countries.

They may not have a nuclear reactor, but they have radiological sources, they have X-ray devices and much more.

This is one of the reasons why we are such strong advocates of regional based training, regional networks and collaboration.

And this is why Australia is committed to working with the IAEA to provide ongoing support to, and collaboration with, the Pacific Island family on nuclear security.

Distinguished Delegates,

The theme of ICONS 2024 is 'shaping the future'.

This is the perfect time for us all to take stock of our existing nuclear security commitments and maximize the benefits of emerging technologies while mitigating the potential risks they may present to nuclear security.

Take AI, for example.

We need to shape relevant AI rules and norms to protect nuclear and related facilities from malicious cyber activity that could compromise nuclear security.

A global effort is required to manage these complex and evolving challenges.

We need your expertise, as the practitioners, the policy makers and the academics.

We need your experience, whether you are in the private sector or in a non-governmental organization (NGO).

We need to work together on shared challenges and respond collectively.

Distinguished Delegates,

Australia's commitment to the Women, Peace and Security (WPS) agenda is steadfast and enduring.

We know that diversity, equity and inclusion are central to establishing and maintaining peace and security.

This applies to nuclear security, where the full, equal and meaningful participation and leadership of women is an essential element.

Australia is proud to partner with Kazakhstan to host a side event at ICONS providing practical insights and actionable guidance on gender equality and inclusive leadership as positive drivers of the future of nuclear security.

The global community is grappling with emerging risks and threats to nuclear security. As they evolve, so must our nuclear security frameworks and responses.

It is good to have so many of you here at ICONS 2024.

Our discussions this week will help shape the IAEA's next Nuclear Security Plan and ultimately advance the global nuclear security agenda.

Strong leadership and shared commitments are essential if we are to keep the world safe and secure from the threat of nuclear terrorism.

Let me conclude by extending Australia's sincere appreciation to you, Co-President, Vice Minister for Energy, Sungat Yessimkhanov, and your entire team, for the combined efforts and collaborative spirit shown during the ICONS 2024 preparatory process. My appreciation also goes to the Programme Committee for their significant efforts in helping to deliver the scientific and technical programme of ICONS.

Last but not least, I extend my personal recognition to the IAEA Director General Rafael Grossi and his team within the IAEA Secretariat for their dedication and tireless efforts to deliver a comprehensive, thought provoking and engaging conference for all of us.

Thank you.

HE Ms Ana Cristina Tinca

Deputy Minister for Foreign Affairs
and State Secretary for Strategic Affairs of Romania

Director General,

Co-Presidents,

Distinguished participants,

It is my pleasure to be here today at the fourth International Conference on Nuclear Security. I would like to thank the Director General of the IAEA for maintaining this topic on the international agenda.

Four years ago, together with Panama, Romania presided over this conference. Including from this perspective, I want to commend our current Co-Presidents, Australia and Kazakhstan, for their contribution to the success of ICONS 2024.

I would also like to convey my gratitude to our distinguished co-chairs of the consultations process, the ambassadors of Australia and Kazakhstan, for their relentless efforts over the past months. Romania subscribes to the Co-Presidents' Joint Statement as the document capturing the determination of the international community to improve nuclear security.

In 2020, we have collectively shown our determination to make significant progress. We agreed upon a wide range of elements essential for further advancing on the path of nuclear security globally. It is in this spirit that we would like to highlight the importance of maintaining these commitments and actions, principles and responsibilities for a safer, more secure world. While ICONS 2020 focused on sustaining and strengthening our efforts, this year's theme focuses on the future.

Distinguished participants,

International security is gravely affected by unprecedented threats and challenges. The Russian Federation's war of aggression against Ukraine is raising nuclear security concerns unseen before. Extensive attacks on civilian energy infrastructure in Ukraine were preceded by the illegal seizure by the Russian forces of the Zaporizhzhya nuclear power plant. These are highly disturbing events and have significantly raised the risk of a nuclear accident, with potential severe consequences for Ukraine and beyond.

Romania would like to commend the IAEA Director General for his work and that of the courageous Agency staff in ensuring that a nuclear incident at Ukraine's nuclear facilities does not take place, while respecting Ukrainian sovereignty and territorial integrity.

Our commitment to Ukraine remains strong, and we fully support the joint statement in this regard.

Amid increasingly volatile risks to nuclear security, including from nuclear terrorism, insider threats, as well as the potential misuse of nuclear material remain significant challenges to international peace and security. The evolving threats encompass a diverse set of actors, potentially accentuated by technological advancements. We strive to advance the state of practice for insider threat mitigation, and we will continue to show our commitment by endorsing the revisions to the Joint Statement on Mitigating Insider Threats, put forward by Belgium and the United States of America.

Distinguished participants,

Although nuclear security is within the responsibility of each Member State, it also requires collaboration. We look towards governments and international organizations coming together to share their expertise and best practices and to facilitate the identification of effective solutions.

The IAEA has proven its central role in coordinating global efforts and assisting Member States implement effective nuclear security. Romania strongly supports the work of the Agency, which continues to provide a valuable contribution to international peace and security. The Agency is indispensable in the development of international security guidance, the provision of support to Member States and the coordination of nuclear security activities.

As a founding member of the IAEA, Romania has extensive knowledge and experience in developing a nuclear programme exclusively for peaceful purposes for more than half a century. As a country known for its high standards and expertise, we are proud of our efforts and advancements and we will continue to strengthen our domestic nuclear safety and security.

In ensuring nuclear security we highlight the need for preparedness and effective response to any potential nuclear incident. As such, the Romanian National Commission for Nuclear Activities Control (CNCAN) has developed a strong collaboration with the IAEA, contributing with technical experts and the active participation of national experts in peer review missions.

Through actions such as the 2023 Integrated Regulatory Review Service (IRRS) mission undertaken by IAEA experts in Romania, we have been able to showcase our systematic approach in addressing the interface of safety and security in the regulatory oversight programme for nuclear power plants. Furthermore, Romania's regulatory infrastructure for nuclear safety and security has been strengthened through the Enhancement of Nuclear Safety, Security and Emergency Preparedness in Romania (NORROM) project. Completed this year with financing from the Norway Grants and supported by the IAEA, this project led to the establishment of a new emergency information and training centre, thus showing our commitment towards preparedness.

Furthermore, our Nuclear and Radioactive Waste Agency, with valuable support through the IAEA technical cooperation programme, commits to strengthen the national infrastructure for the safe and secure management of radioactive waste.

We also operate a Nuclear Forensics Laboratory, within the auspices of the Horia Hulubei National Institute for Research and Development in Physics and Nuclear Engineering. This has allowed for the development of a unified nuclear forensics approach as a tool for the growth of nuclear security. This is an opportunity to facilitate connections and mutual nuclear forensics assistance and information sharing between countries.

Excellencies,

Distinguished colleagues,

International cooperation stands at the basis of technological progress in the field of nuclear energy. By working with the United States of America and other key partners, Romania is determined to develop its nuclear energy programme through enhancing large scale production capacity and deploying the innovative small modular nuclear reactor technology.

Within our development process, we are committed to ensuring nuclear safety and security during the design, development and deployment phases. To this end, Romania is actively participating in the Nuclear Harmonization and Standardization Initiative (NHSI), launched by the IAEA Director General.

At the national level, we put an emphasis on the early stages of SMR development, and we are keen to ensure that advancements in civil nuclear energy technologies will bring a positive contribution to achieving our climate goals. Our support is evidenced not only by our actions at the national level, but also through the support we lend to the IAEA Atoms4NetZero initiative.

It is also our view that a strong nuclear security regime remains a critical element for the advancement of nuclear energy globally and will continue to take actions described in the Joint Statement on Advanced Nuclear Energy Technologies.

In concluding, allow me to reassure you of Romania's determination for continuous strengthening of nuclear security. We aim to uphold our commitments and contribute to furthering the achievements made, by fostering cooperation with international partners and continuing our relentless support to the work of the IAEA.

Thank you.

2.5.   OUTGOING CONFERENCE PRESIDENT, PANAMA

Statement translated from Spanish.

HE Mr Dario Chiru Ochoa

Ambassador and Permanent Representative of Panama

His Excellency Tim Watts, Assistant Minister for Foreign Affairs of Australia,

His Excellency Sungat Yessimkhanov, Vice Minister of Energy of Kazakhstan,

Mr Rafael Mariano Grossi, Director General of the International Atomic Energy Agency,

Heads of Delegations,

Ambassadors,

All delegations,

Let me first extend my sincere congratulations to the two Co-Presidents of ICONS 2024 who have masterfully led our preparatory work, as well as to the Director General and the Secretariat for their efficient and methodical support for the preparation and organization of this important meeting.

As the outgoing Co-President of ICONS 2020, together with our friends from Romania, I am pleased to note the significant achievements of ICONS 2020, which marked a milestone in global efforts to continue promoting global awareness and commitment to the effective and comprehensive strengthening of nuclear security for all nuclear facilities and material.

The 2020 Conference made it clear that each State is responsible for assuming full responsibility for nuclear security at its borders, in accordance with its international obligations and respective national legislation, while the IAEA was in charge of continuing to assume its crucial role of coordinating and providing technical advice in this area with a view to making available, upon request, the necessary assistance to fulfil its responsibility.

Distinguished delegations,

The complex international situation we face today has renewed the importance of redoubling our efforts to strengthen our structures and policies in the area of nuclear security, non-proliferation and disarmament.

In this new edition of ICONS, nuclear security faces growing challenges and emerging threats, given the development and proliferation of innovative technologies, including artificial intelligence and the accelerated production of small or modular reactors.

However, emerging technologies also offer new opportunities to strengthen our capabilities with a view to guaranteeing nuclear security for all. My delegation has no doubt that this great conference will be able to overcome the serious challenges we face, and will take decisive steps towards a safer world in which nuclear energy and nuclear technologies and applications are only used for the progress of peoples.

My delegation wishes you the best that this ICONS 2024 may rise above the circumstances to leave us with guidelines and commitments that contribute to advancing the consolidation of nuclear security, overcoming obstacles for the common good.

Thank you very much.

3. CLOSING ADDRESSES

3.1. CONFERENCE CO-PRESIDENT, AUSTRALIA

HE Mr Ian Biggs

Ambassador and Permanent Representative of Australia

Director General Grossi,

Distinguished Co-President from Kazakhstan,

Distinguished Delegates,

We are at the end of ICONS 2024, a significant milestone for our global efforts to strengthen nuclear security worldwide — the latest but by no means the last stage in the global collaboration that has been expanding since at least the Nuclear Security Summits from 2010.

If there is one thing this week has highlighted, it is that nuclear security cannot be achieved unilaterally. Governments, international organizations, industry, civil society and the next generation of nuclear security professionals must work together — and have been working together so impressively, this week.

The active and inclusive participation of almost all IAEA Member States at ICONS 2024 —from ministers, policy makers and diplomats to experts and industry representatives — is what makes this international gathering so very special, and has been key to our productive, thought provoking and forward looking deliberations this week.

On behalf of Australia, entrusted with this intimidating role — in recognition, I would like to think, of our own nuclear security record — I thank all IAEA Member States for their attendance and engagement. ICONS is a success because of each of you.

Distinguished Delegates,

One of our objectives and duties as Co-Presidents of ICONS 2024 was to deliver an ambitious, forward leaning Ministerial Declaration. On behalf of Australia, I wish to once again acknowledge the sustained efforts of Member States to help us to arrive at a consensus Ministerial Declaration over the past three months.

It is disappointing that despite being so very close to reaching consensus, we were ultimately unable to achieve agreement. Nevertheless, the Co-Presidents' Joint Statement — which reflects a compromise agreed to by the majority of us here today — and already enjoys the declared alignment of over 50 Member States, captures our collective political will and commitment to further strengthen nuclear security, including in response to emerging and developing technologies and new nuclear security challenges.

We continue to encourage Member States to align with our Co-Presidents' Joint Statement over the coming weeks, and to translate its contents into practical actions that will make the world secure from the threat of nuclear terrorism, and more prosperous and safer for our communities to thrive.

And now let me turn to my distinguished Co-President and dear friend from Kazakhstan, Ambassador Tileuberdi. It has been a privilege working with you, Yerzhan, and Abylaikhan over the course of the last four months in preparation for ICONS 2024. Your dedication and tireless efforts in working with Australia — with Marina, Katie and Jonathon — as a seamless team is sincerely appreciated.

My profound appreciation also goes to the ICONS Programme Committee, and to the Delegations of Costa Rica and Sweden for their leadership in this regard. Your efforts resulted in a thought provoking scientific and technical programme that has been essential for facilitating knowledge exchange amongst nuclear security practitioners. Nuclear security is a heavy responsibility for every government, and the extraordinary assemblage of exhibitions, side events, scientific presentations, and human expertise in Vienna this week is testimony to the seriousness and determination with which all of us are undertaking that responsibility.

Last but not least, I extend my personal recognition to the IAEA Director General, Rafael Grossi, and his team within the Secretariat for their dedication and efforts to deliver this successful conference. I recognize that a vast number of Agency staff from numerous departments have been working tirelessly over many, many months. Australia is sincerely thankful for all of your individual and meaningful contributions.

Despite my desire to personally recognize and name each of you, I'm conscious that my very long list of people to thank would prevent many of you from catching your flights home.

However, it would be churlish of me not to acknowledge, in particular and publicly, the efforts of Director Elena Buglova, and the relentless work of the Scientific Secretary, Sara Mroz and her team members Christian Deura and Bryan Denehy. Recognition must also go to Sanjai Padmanabhan and the entire conference services team. Thank you.

And with that, I wish to thank all delegates once again for making this conference so memorable. I am confident that the knowledge, insights and relationships we have all gained this week will help us to shape the future of nuclear security, globally. We shall meet again in this format and, one hopes, at ministerial level, four years from now.

Thank you all, and I wish you a safe return home.

HE Mr Mukhtar Tileuberdi

Ambassador and Permanent Representative of Kazakhstan

Director General Grossi,

Co-President Ambassador Biggs,

Excellencies,

Distinguished delegates,

It is my privilege to mark today the successful conclusion of the fourth International Conference on Nuclear Security – ICONS 2024.

I am very grateful to Ambassador Ian Biggs and his professional team for their utmost dedication and joint efforts in preparing the ministerial segment of the conference. Even though the distance between Astana and Canberra is more than 12 000 km, and they are two completely different parts of the world, the spirit of the ICONS conference has united us like never before. We were united in our sincere hope and tireless efforts to make this conference successful and effective in consolidating the international community around the common goal of shaping a better future in nuclear security.

In this regard, I commend the personal efforts of the Director General Rafael Mariano Grossi and the IAEA team, in particular, the Division of Nuclear Security and IAEA Conference Services, for providing us with this wonderful opportunity and guiding us throughout the whole process.

My sincere thanks also go to the co-chairs of the Programme Committee for their skilful leadership in preparing the scientific and technical programme of ICONS 2024, as well as to all delegations for their active participation.

Distinguished delegates, ICONS 2024 has indeed brought together ministers, policy makers and nuclear security experts from 142 countries who are committed to shaping our better future in such an important field as nuclear security.

Ministers and high representatives from Member States once again reaffirmed that the establishment of robust national nuclear security regimes, development of predictable advanced technologies and universalization of core legal instruments in the field of nuclear security and safety will ensure a guaranteed nuclear energy chain for humankind and promote the implementation of sustainable development initiatives around the globe. Enhanced cooperation and information sharing among countries will help prevent illicit trafficking of nuclear material and technologies. In this regard, important forums such as ICONS, under the auspices of the Agency, will continue to build momentum for strengthening partnerships among States, sharing best practices and further reinforcing the role of the IAEA in nuclear security.

Dear friends,

It is very unfortunate that we could not overcome our differences to achieve a consensus based Ministerial Declaration for the conference. Both Co-Presidents put their best efforts and energy to ensure an inclusive and transparent negotiating process in a very friendly atmosphere. Even though we ended up adopting the Co-Presidents' Statement, this document carries the weight of our shared commitments, principles and priorities. I thank delegations for aligning themselves with this statement, and I call on Member States to use this statement in our further discussions on nuclear security. The Joint Statement of the Co-Presidents contains important messages and outlines the latest trends in nuclear energy and global security. This includes ensuring gender equality in a traditionally male dominated industry such as nuclear energy.

Kazakhstan upholds the principle of gender equality in its domestic policy and has taken firm systematic steps to strengthen women's rights and opportunities. My President, Kassym-Jomart Tokayev, recently signed a new law aimed at eradicating gender based violence and promoting gender equality. We believe it's crucial to increase the participation of women in all sectors of business and politics. Diverse teams bring diverse perspectives, they foster creativity and innovation. In this regard, it was a pleasure, with my Co-President from Australia, to co-host a side event on gender equality and women's empowerment, and to share with you some compelling facts on women's representation in the nuclear sector and to set more ambitious goals in this area.

Distinguished Delegates,

Thank you once again for your time and active participation in this fascinating week-long conference on nuclear security. I firmly believe that through our joint efforts, we can ensure that nuclear technologies are used safely for the benefits of humankind and shape a better future for generations to come. Kazakhstan attaches great importance to the international principles of cooperation in the energy sector and reaffirms its commitment to strengthening global energy security. I would like to wish all the participants who have come from their capitals and other places, a safe journey back to their homes and all the best in their noble missions.

And I thank you.

Mr Rafael Mariano Grossi

Director General
International Atomic Energy Agency

We come to the end of what has been a very successful week indeed. It has been said, not only by the Co-Presidents, and many of you, also the young representatives here, reminding us that they are going to be taking over from us and leading the world in its efforts in the area of nuclear security and I think this is so important.

I will not repeat what has been said. Many good ideas, many topics have been dealt with in the context of the deliberations here. More than 2000 participants have gathered together and it's very encouraging to see that at this closing session on a Friday afternoon, this hall is as packed as on the opening on Monday morning. This is an indication of commitment, of interest, which is so important. I believe that the ideas that have been gathered in the very important process that led us to the statement made by the Co-Presidents and the many presentations that we have had, indicate one thing: nuclear security is an issue that — above and beyond circumstantial political differences that sometimes and quite often have very little to do with the essence of nuclear security — is shared by all of you, by all of us.

Most importantly, I would say, these regular ICONS conferences that are so important for us to take the temperature of the global impressions, to gather in a very significant and high level way, periodically every four years, do not detract from one very important thing that I would like us to be reminded of: the fact that when the lights go off and when delegates go back home and turn their professional attention to other things, the work on nuclear security continues. It never stops.

On Monday there will be people working on nuclear security. There will be delegations of the IAEA taking off from Vienna to perform a security advisory assessment or a review mission. There will be people coming to our fantastic Nuclear Security Training and Demonstration Centre. The work, the real work on nuclear security, never stops and this I think should be a cause for satisfaction. At this time, as we said just a few days ago, when also the world, our societies, our countries, are turning their eyes to nuclear to get solutions for their many problems and challenges that we have, be it through the production of CO_2 free energy or cancer care or food security for those who need it. Nuclear security is an integral part of that, so let's celebrate this consensus which exists above and beyond the words of diplomatic papers. It's in the real world.

Nuclear security is there, will be there and it's not going anywhere.

Thank you very much. And with this, we close ICONS 2024.

4. CO-PRESIDENTS' REPORT

4.1. INTRODUCTION

The 2024 International Conference on Nuclear Security: Shaping the Future (ICONS 2024) was convened by the IAEA at its Headquarters in Vienna from 20 to 24 May 2024.

The purpose of the conference was to provide a forum for ministers, policy makers, senior officials and nuclear security experts to discuss the future of nuclear security worldwide, whilst providing opportunities for exchanging information, sharing best practices and fostering international cooperation. The conference succeeded in achieving the following:

- Raising awareness on a wide range of nuclear security topics to maintain and further strengthen national nuclear security regimes, as well as underscoring the critical role of international cooperation to strengthen nuclear security globally;
- Reviewing national and international nuclear security experiences and achievements along with current approaches and trends, and highlighting areas that may need more focused attention, including in the technological domain;
- Promoting IAEA nuclear security guidance, and the experience of States in their application, including through peer reviews, advisory services and capacity building;
- Promoting the sharing of information and good practices in nuclear security while protecting sensitive information;
- Reaffirming and supporting the central role of the IAEA in strengthening nuclear security globally and serving as a convening force for international activities in the field of nuclear security;
- Highlighting and promoting the IAEA's efforts to promote adherence to relevant international legally binding instruments and commitment to the legally non-binding instruments;
- Demonstrating the importance of recruiting and fostering the development of the next generation of nuclear security professionals;
- Discussing further enhancements of IAEA nuclear security activities and their sustainability;
- Stressing the critical role that nuclear security plays for all countries;
- Emphasizing the fundamental enabling role that nuclear security plays in the global pursuit of the United Nations 2030 Agenda for Sustainable Development;
- Orienting the international nuclear security community towards the threats and challenges which lie ahead in order that its members can shape a future that is safe, secure, and sustainable.

The conference did not discuss any sensitive nuclear security information.

Over 2000 participants from 142 countries and 16 invited organizations registered to participate at ICONS 2024, with 11 ministers, 24 vice-ministers and 14 other high ranking officials participating in the ministerial segment of the conference.

4.2. MINISTERIAL SESSION

4.2.1. Opening plenary session

During the opening plenary session of ICONS 2024, IAEA Director General Mr Rafael Mariano Grossi and the conference Co-Presidents, Mr Tim Watts MP, Assistant Minister for Foreign Affairs of Australia, and Mr Sungat Yessimkhanov, Vice Minister of Energy of Kazakhstan, delivered opening remarks.

Following the remarks delivered by the Director General and Co-Presidents, Ms Ana Cristina Tinca, Deputy Minister for Foreign Affairs and Secretary of State for Strategic Affairs of Romania, and Mr Dario Chiru Ochoa, Ambassador and Permanent Representative of Panama, both provided remarks in their capacities as outgoing Co-Presidents of ICONS 2020.[4]

The Co-Presidents then officially opened the conference and presented a Co-Presidents' Joint Statement in lieu of a Ministerial Declaration on Nuclear Security. While many Member States agreed on the tenets of a draft Ministerial Declaration that was negotiated in the months preceding the conference, consensus was not achieved. The Co-Presidents' Joint Statement sought to reflect the outcome of the many negotiations and open-ended working groups, highlighting the critical role that nuclear security plays in achieving global peace and security and reaffirming the central role of the IAEA in coordinating international nuclear security activities. The Co-Presidents requested that Member States indicate their alignment with the Joint Statement during their national remarks.[5]

4.2.2. National statements

A total of 99 national statements were delivered by ministers and other heads of delegation, along with 3 joint statements, 2 statements delivered by international organizations, and 1 country exercising its right of reply.

4.2.3. Plenary panel: Securing sustainable progress — The important role of nuclear security in advancing the sustainable development goals

One plenary panel was held during the ministerial segment of the conference. The session sought to highlight the ways in which nuclear security underpins the global pursuit of the United Nations' 2030 Agenda for Sustainable Development, with particular focus on Sustainable Development Goals (SDGs) 2 (Zero Hunger) and 3 (Good Health and Well Being). Panellists provided examples of the ways in which nuclear security enables various peaceful applications of nuclear technology, including using nuclear techniques to improve crop yields, promote environmental sustainability, and ultimately work towards improving food security (SDG 2). The IAEA's Atoms4Food programme (implemented jointly with the Food and Agriculture Organization of the United Nations) was specifically addressed. Panellists also discussed the integral role of nuclear security as the Agency works to improve States' capacities for cancer therapy within the framework of the IAEA's flagship cancer initiative 'Rays of Hope' (SDG 3). The discussion highlighted the critical role of nuclear security education to ensure that individuals who support activities that utilize nuclear technologies are vigilant and adequately prepared to secure radioactive and nuclear material from malicious actors.

4.2.4. Ministerial segment event: Beyond borders — A collaborative discourse on the future of nuclear security

The Secretariat also organized an event outside the Vienna International Centre for ministers and other heads of delegation. This included a round table discussion during which the panellists emphasized that, given the expanding interest in nuclear technologies for a variety of peaceful applications, the role of nuclear security in the international landscape will grow in importance. Moreover, it was noted that in today's environment, potential threats have become increasingly international in nature. In order to effectively deal with this shift, panellists concurred that international cooperation and strong partnerships will be essential. They stressed the need for nuclear security experts to translate the technical language of nuclear security into ideas that are digestible for wider audiences, to include lawmakers and parliamentarians, in order that decision makers at all levels can better understand the threats that exist and the steps that can be taken to mitigate them.

---

[4] All opening addresses are presented in Section 2 of this publication.
[5] The statement by the Co-Presidents is presented in Section 5 of this publication.

4.2.5. Interactive ministerial session

Ministers and other heads of delegation were invited to a scenario-based, interactive policy discussion organized by the Secretariat. The ICONS moderator, acting as a national security briefer, walked participants through two separate fictional nuclear security event scenarios. Participants were asked to respond to question prompts following the presentation of each scenario, after which the moderator facilitated discussion about submitted responses. The main foci of the session were the importance of the universalization of the Convention on the Physical Protection of Nuclear Material (CPPNM) and its Amendment (A/CPPNM) and political commitments to the Code of Conduct on the Safety and Security of Radioactive Sources. Discussions also addressed the requirements and resources needed for States to effectively prevent, detect and respond to nuclear security events.

4.3. SCIENTIFIC AND TECHNICAL SEGMENT

4.3.1. Opening plenary session

The scientific and technical segment was opened on the second day of the conference and ran in parallel with the second day of the ministerial segment. During the opening plenary session for the scientific and technical segment, remarks were delivered by Deputy Director General and Head of the Department of Nuclear Safety and Security, Ms Lydie Evrard; by the Director of the Division of Nuclear Security, Ms Elena Buglova; and by the Scientific Secretary of ICONS 2024, Ms Sara Mroz.

During these remarks, the speakers acknowledged the efforts of the Programme Committee, co-chaired by Costa Rica and Sweden, and which consisted of representatives from 45 Member States and 5 international organizations. An overview of the conference programme was presented, and it was noted that the Secretariat made concerted efforts to ensure the diversity of presenters, panellists, session chairs and participants. Speakers introduced the four main themes of the conference under which the 52 technical sessions were organized. The themes of the conference were: Policy, law and regulations for nuclear security; Technology and infrastructure for nuclear security prevention, detection and response; Capacity Building for nuclear security; and Cross-cutting nuclear security topics.

Following the introductory remarks, the Nuclear Security Delegation for the Future was introduced by the Secretariat. This cohort of young nuclear security professionals included 24 participants from 19 countries selected ahead of ICONS to take an active role in the conference and be provided with various professional development opportunities.

4.4. POLICY, LAW AND REGULATIONS FOR NUCLEAR SECURITY

Technical sessions organized under this theme addressed various dimensions of national and international policy, law and regulatory frameworks to reinforce and bolster the global nuclear security regime. Special attention was given to the ways in which the international community could adapt frameworks to address emerging challenges, while also communicating to all States the importance of signing onto international legal instruments that reinforce international norms associated with nuclear security (e.g. the A/CPPNM).

4.4.1. Plenary panel: Policy, law and regulations in an evolving nuclear security landscape

The first plenary panel of the scientific and technical segment of the conference explored the challenges associated with developing, implementing and adapting policy, law and regulatory frameworks in a rapidly changing international landscape. Panellists also identified potential gaps and areas for improvement in developing and implementing international legal and regulatory frameworks, while emphasizing the importance of identifying sustainable solutions for strengthening nuclear security in an environment of uncertainty. All participants stressed the importance of international instruments such as the A/CPPNM and the International Convention for the Suppression of Acts of Nuclear Terrorism (ICSANT), and the critical need for Member States to sign on to these instruments. Additionally, panellists addressed national perspectives on developing and implementing regulatory frameworks for new technologies such as small modular reactors (SMRs), while concurrently commenting on industry's role in this process. The participants recognized the continuous work being done by the IAEA to assist Member States in enhancing their legislative and regulatory capacities for robust national nuclear security regimes. All those in attendance were strongly encouraged to take advantage of these services and to advocate for the universalization of the A/CPPNM.

4.4.2. Technical sessions

4.4.2.1. *Global perspectives on nuclear security regulations for SMRs*

This technical session provided an overview of different perspectives on developing and implementing regulations for SMRs given the differences between this reactor technology and traditional large reactors and the potential impact on standing regulatory frameworks. Several panellists highlighted the importance of security by design approaches and the consideration of interfaces between security, safety and safeguards. Various regulatory approaches were discussed, to include prescriptive, performance based and combined approaches, along with multiple elements of a given State's regulatory infrastructure, to include licensing, design and change management, and computer security. Panellists from North America, Europe and Africa each acknowledged the unique challenges that these technologies pose for their specific regions, while recognizing that there were lessons to be learned from diverse national experiences.

4.4.2.2. *Strategies for strengthening regulatory frameworks for the transport of nuclear and other radioactive material*

During this technical session, panellists provided diverse perspectives on their respective national approaches to implementing and improving regulatory frameworks for the transport of nuclear and other radioactive material. One panellist acknowledged that given the increase in the use of these technologies for peaceful purposes, there has been a need to adapt their national strategy to the increase in transport activities. Their national regulatory authority has collaborated extensively with other national competent authorities to improve legal frameworks and will continue to foster these cooperative relationships. Another panellist detailed some national experiences with utilizing IAEA guidance to develop national regulatory frameworks for transport, suggesting that there may be value in the development of an integrated guideline for the safety and security of nuclear and other radioactive material transport. There was also a presentation about the unique challenges posed by trans-shipment platforms, along with some of the steps taken within the presenter's country to mitigate potential issues.

4.4.2.3. *A continental case study: Lessons learned in regulation development in the African region*

This technical session discussed the development and implementation of legislative and regulatory frameworks for nuclear security in several countries of the African region, to include representatives from Cameroon, Ghana, Kenya, Nigeria, Tanzania and Uganda. Panellists gave status updates about their States' alignment with IAEA safety standards and nuclear security guidance, while detailing the value gained from their respective participation in IAEA hosted consultancy meetings, workshops and staff training. Panellists discussed some of the challenges associated with coordinating between numerous national stakeholders when developing regulations and some of the methodologies that can be implemented to ensure all viewpoints are considered. One panellist addressed national experiences in amending existing regulations given the potential expansion of new technologies in the region (i.e. SMRs). Another detailed their national experience of interfacing with neighbouring countries to develop cooperative approaches to nuclear security, specifically in the development of memorandums of understanding for the import and export control of radioactive material at borders, and for the establishment of joint measures at one-stop border posts.

*4.4.2.4. Charting the way forward: Assessing current and future challenges on the implementation
of the A/CPPNM*

This technical session focused on various aspects of the A/CPPNM. Key presentations included strategic steps suggested for the second Conference of the Parties to the A/CPPNM, the role of a repository of national legislations in combating nuclear terrorism, and the use of effective self-assessment tools to enhance nuclear security. Additionally, participants covered legal and regulatory challenges, methodologies for evaluating the adequacy of the A/CPPNM, and the critical need for inclusivity and targeted support for smaller States. The session concluded with a discussion addressing best practices, the importance of self-assessment and peer reviews, and recommendations for enhancing international cooperation. Panellists agreed that collaborating with international partners was crucial to addressing emerging threats and pursuing harmonization of relevant legal matters to strengthen nuclear security worldwide. It was suggested that a framework be established for the regular review and adaptation of the A/CPPNM to ensure that it responds to contemporary and future security challenges.

4.4.2.5. Better together: Experiences in international cooperation for regulatory implementation

During this technical session, panellists discussed successful experiences in strengthening national nuclear security regimes through international cooperation and collaboration. Presenters noted that international cooperation needs to be driven by national needs, and acknowledged the central role that organizations like the IAEA play in these efforts. Panellists also recognized how regional networks and NGOs (e.g. the African Commission on Nuclear Energy, and the African Center for Science and International Security) can supplement the work of international organizations, while also discussing experiences in bolstering national nuclear security regimes through bilateral relationships. One panellist noted that nuclear maritime systems pose potential challenges and suggested that the IAEA should work closely with the International Maritime Organization to harmonize nuclear security requirements for these systems. Participants recognized that all countries, regardless of the maturity of their nuclear security programmes, can contribute to strengthening nuclear security globally, while encouraging those with advanced programmes to commit to being resource providers for the international community.

4.4.2.6. Bits and bytes: Computer security considerations for regulatory frameworks

This session provided a comprehensive overview of ways to develop and implement computer security regulations for nuclear facilities. Several key principles were highlighted, to include implementing risk informed and defence in depth approaches, along with the need to strike a balance between safety and security. Panellists also noted that consistency between computer security regulations and other national regulations is important. Presentations pointed to the ways in which computer security regulations might be tailored for different operational settings, as one presenter discussed the steps being taken to draft regulations for the computer security of advanced reactors, and another discussed the ways in which computer security could be strengthened at radioactive material facilities. It was noted that in order to develop strong computer security regulations, staff of regulatory authorities need to have professional competence in these areas. Cooperative, bilateral training programmes can help to fill these gaps, along with the services offered by the IAEA.

4.4.2.7. Strengthening regulatory frameworks for radioactive sources throughout their life cycle

This session highlighted the ways in which regulatory frameworks, complemented by industry standards, can be tailored and implemented to address challenges associated with handling radioactive sources throughout the life cycle. Presenters discussed national experiences with developing new regulatory frameworks, while also addressing national and regional challenges associated with regulating the transnational movement of radioactive sources. One panellist presented a new standard developed by the International Electrotechnical Commission for the security of medical electrical equipment using high activity radioactive sources, recognizing that although the IAEA and industry standards have different approaches, these efforts are complementary in nature and lead to a strengthened impact. Novel approaches to managing radioactive and disused sealed radioactive sources were also discussed, including a new open source online database for source security frameworks, and a first of its kind borehole disposal project in Malaysia.

4.4.2.8. Regulatory frameworks around the world: National best practices

During this technical session, a diverse set of panellists shared varying experiences with developing new regulatory frameworks and amending existing ones to ensure robust national nuclear security regimes. Panellists noted that open, cooperative approaches between various stakeholders within the State are essential. Presenters discussed the need for regulators to develop open lines of communication with operators, along with national experiences in developing violation prevention programmes. One presenter discussed the possibility of leveraging potential interfaces between a State's established emergency preparedness and response framework and a nascent nuclear security framework. Discussions touched on numerous facets of developing robust national nuclear security regimes, to include experiences with developing an Integrated Nuclear Security Sustainability Plan (INSSP) and ratifying international nuclear security instruments. It was noted that prolonged regulatory drafting and promulgation processes can be problematic and measures were needed to ensure efficient approaches.

4.4.2.9. From commitments to action: Perspectives on implementing legal provisions for nuclear security

This technical session discussed numerous topics related to the implementation of legal frameworks for nuclear security, along with associated challenges. Key themes included the critical role of international cooperation, particularly through conventions such as the ICSANT, along with the need for a dynamic and responsive nuclear security regime to effectively address evolving threats. The session also highlighted practical insights about how States can meet international nuclear security obligations and explored the relationship between the Treaty on the Non-Proliferation of Nuclear Weapons (NPT) and nuclear security, underscoring the need for global collaboration. Panellists noted that integrating international standards into national frameworks and updating national laws to address new threats was an essential process for robust regimes. Panellists stressed the importance of assisting developing countries with these processes through technical assistance and capacity building programmes, which will in turn reinforce global nuclear security.

4.4.2.10. Regulations for the future: Adapting and implementing regulatory frameworks for material and facilities

Panellists during this session covered various perspectives on how the security of nuclear material and facilities are addressed in national regulatory frameworks. One panellist discussed the possibility of expanding guidance to operators on how to implement a design basis threat (DBT), with practical considerations for the development of location specific DBT scenarios and how to assess them. Another panellist presented the ways in which the nuclear security regulatory framework was revised in their State through the use of performance based approaches, while considering various challenges associated with SMRs, computer security, and the safety–security interface. Another presentation detailed a novel 'fitness for duty' programme for advanced reactors to ensure their sustained secure operation. Other speakers discussed radioactive source safety and security at a gamma irradiation facility, and the evolution of regulations to account for the need to manage changes resulting from industrial activity or enhancement programmes at nuclear facilities. It was noted that IAEA guidance on DBT for regulators is available in IAEA Nuclear Security Series No. 10-G, National Nuclear Security Threat Assessment, Design Basis Threats and Representative Threat Statements, while guidance for operators is not. It was suggested that the IAEA might consider developing guidance for the use of DBT by operators.

4.5. TECHNOLOGY AND INFRASTRUCTURE FOR NUCLEAR SECURITY PREVENTION, DETECTION AND RESPONSE

The majority of the technical sessions which comprised the conference programme fell under the theme of technology and infrastructure for nuclear security prevention, detection and response. Given the pace at which the international environment is evolving, sessions organized under this theme sought to address a variety of emerging threats while working towards collective approaches through the sharing of novel research, national experiences and new perspectives on emerging technology.

4.5.1. Plenary panel: Managing the threats and benefits of emerging technologies

This plenary panel included a diverse set of experts who each contributed a unique regional perspective on the challenges associated with emerging technologies. One of the drivers of the discussion was the role that artificial intelligence (AI) might play moving forward. Concerns were expressed about removing humans from decision making processes without fully understanding the implications of this technology's practical use. The panel shifted the discussion to the growing interest in SMRs, and the need to assist international partners with their secure deployment. Panellists discussed the benefits of both vendors transparently demonstrating that security by design has been applied to their SMRs and purchasers making market demands for demonstrable security by design. The panel concurred that technical assistance projects and knowledge sharing will be essential to ensure that new technologies will be regulated and employed in a responsible and secure manner. Similarly, training will play a key role in ensuring that practitioners are equipped with the tools and knowledge to stay apprised of rapid developments in the field. Panellists noted that utilizing capabilities such as IAEA's Nuclear Security Training and Demonstration Centre will play a central role in ensuring that practitioners are ready to confront the challenges of the future.

4.5.2. Technical sessions

4.5.2.1. *Exploring the interfaces between nuclear security and State systems of accounting and control*

This technical session explored the nexus between those areas. Panellists acknowledged that centralized nuclear material accountancy is an effective tool to detect theft or repeated diversion of small quantities of materials; contributes to the management of crisis situations; and enables States to comply with the provisions of their safeguards agreements with the IAEA, and meet any relevant nuclear security requirements. Several challenges associated with reliable State systems of accounting and control were noted, including the possibility of limited resources or skilled analysts, potentially inadequate national regulatory frameworks, and a lack of globally acceptable guidelines. One panellist detailed the ways in which nuclear material accountancy and control can serve as a strong deterrent against insider threats, in that it delivers the ability to maintain precise knowledge on nuclear material attributes and locations, while ensuring the activities performed in connection with these materials have been properly authorized. Panellists stressed the importance of close collaboration between different multidisciplinary subject matter experts through consultancy meetings, workshops and reviews of previous lessons learned.

4.5.2.2. *Protecting the public: Implementing nuclear security measures for major public events*

During this technical session, panellists discussed measures to incorporate nuclear security considerations into planning for major public events. Several case studies were presented, including national experiences with planning the Commonwealth Heads of Government meeting in Rwanda, the 50th World Petanque Championship in Benin, COP 28 in the United Arab Emirates and the U-20 Women's World Cup in Costa Rica. Panellists highlighted the benefits of clear and consistent coordination between multiple stakeholders, including their national governments, and the importance of providing effective training and timely preparation. Pre-event, during-event, and response phases can all present challenges and need adequate contingency plans in place. Other lessons learned included the need to clearly define roles and responsibilities, the importance of adequately preparing and testing detection instruments, and the value of conducting preparatory exercises in the lead up to the event. One panellist noted that timely, accurate and clear public communication is fundamental to major public events, and that there need to be strategies in place prior to event execution. Several acknowledged the extensive help and support received from the IAEA, and the value of the training courses offered by the Agency.

4.5.2.3. *Nuclear forensics' role in bolstering international nuclear security*

This technical session focused on various aspects of national nuclear forensics programmes, along with means and methods to enable practitioners or strengthen national programmes. Panellists discussed various national and international training initiatives, including one State's experience with instituting monthly drills and another State's experience with vocational training, e-training, field exercises and the IAEA International Training Course on Nuclear Forensics Methodologies. Other panellists discussed lessons learned from the development of national nuclear forensics databases, along with experiences in establishing partnerships with entities external to the State. The session concluded that, to develop and expand nuclear forensics capabilities globally, contributors need to develop the science behind and better understand nuclear forensic signatures, while continuing to foster collaboration between scientists, investigators and policy makers. It was stressed that a collaborative global effort is needed in order to continue making advances in the field of nuclear forensics.

4.5.2.4. *Improving national detection and response capabilities for material outside of regulatory control*

During this technical session, panellists presented on several projects and capabilities that may help to improve efforts to detect nuclear or other radioactive material out of regulatory control. Panellists stressed the need to strengthen national regulatory controls, provide continuous technical expert support, and work towards enhancing the software and technology of existing security systems. Presentations ranged from national case studies on enforcing regulatory frameworks to the real world testing and application of new technologies for use by practitioners. These new technologies included a newly developed gateway radiation monitoring system; advanced machine learning algorithms to recognize innocent radiation and reduce nuisance alarms; and the IAEA Mobile-Integrated Nuclear Security Network (M-INSN). The session highlighted the power of data to improve the operation of existing detection systems and ultimately alleviate the burden on front line officers. It was suggested that the IAEA continue to promote the use of open source software (e.g. M-INSN) that users in Member States can adapt to their needs.

4.5.2.5. *Secure from the start: Issues in supply chain security*

During this session, panellists addressed the critical issue of computer security in nuclear supply chains, highlighting vulnerability risks due to reliance on third party solutions and emphasizing the need for robust, tailored computer security frameworks. They underscored the complexity of these challenges and the importance of collaborative approaches, continuous improvement and maturity models for effective security management. Key recommendations included enhancing stakeholder engagement, leveraging technology, and ensuring adequate resources and coordination to protect against cyber-attacks. Panellists noted that computer security maturity models and supply chain models are available online and provide valuable resources for establishing regulatory standards that guide actions. The IAEA non-serial publication, Computer Security Approaches to Reduce Cyber Risks in the Nuclear Supply Chain, was also recommended as a vital resource that can be downloaded from the IAEA website. Further emphasis was placed on the importance of cooperation between facilities and vendors to align expectations throughout the supply chain.

4.5.2.6. *From construction to closure: Approaches to sustainable and secure lifetime management and decommissioning practices*

This technical session emphasized that nuclear security should be a priority throughout the lifetime of nuclear facilities in order to enable the safe, secure, and sustainable use of nuclear energy, science and technology. Panellists focused on several real world decommissioning case studies of nuclear power plants, along with the decommissioning of a reprocessing plant. Various aspects of these projects were presented, including the development of a high activity solid waste measurement system; the evolution of national regulatory standards for decommissioning; and the institution of a graded approach to physical security, computer security and 'fitness for duty' at decommissioning facilities. Panellists agreed that robust security measures are crucial in the design, construction, operation and eventual decommissioning of reactors, and acknowledged that the decommissioning of nuclear reactors poses a unique set of challenges that may be addressed early in the design process. One presenter discussed lessons from a decommissioning project in their country, in which assistance was provided by an international partner. It was noted that States can benefit from international cooperation and that sharing lessons from such projects can be enormously helpful during decommissioning phases.

4.5.2.7. *Exploring the practical uses and potential threats of artificial intelligence*

This technical session focused on the emerging applications of AI to nuclear security, while concurrently addressing the potential challenges it may pose to the field. Panellists discussed AI as a potential tool to enhance various operations, to include examining fuel elements for damage, managing knowledge, enhancing robotics, providing nuclear emergency response recommendations, and deterring intrusions by malicious actors. One panellist discussed the potential for AI to be used to monitor for 'red flags' in personnel, as well as for system anomalies, to better identify potential insider threats. Panellists also acknowledged concerns that AI is developing rapidly, which could lead to it being entrusted with increasingly vital tasks without full understanding of the underlying risks or vulnerabilities. Despite these concerns, there was general agreement that AI will prove to be a net positive for nuclear security, but it will need advocates to bridge the divides between AI experts, policy makers and the broader nuclear security community.

4.5.2.8. *Is the source secure? Case studies in radioactive source security*

During this technical session, panellists discussed the security of radioactive sources in their respective States. Presentations covered diverse topics, including physical protection design and evaluation at facilities with radioactive sources; the successful replacement of high activity radioactive sources with alternative technology through international cooperation programmes; collaboration between States to remove disused high activity ^{60}Co sources from hospitals; and positive outcomes of cooperation between federal agencies in a task force on radiation source protection and security. Panellists highlighted the importance of international cooperation and assistance; the need for effective coordination among multiple national stakeholders; and the importance of continuous self-assessments. Panellists stressed that when a State pursues alternative technology, stakeholders need to be well informed, and establish a national strategy for the management of disused radioactive sources while concurrently leveraging international partnerships, if assistance is needed.

4.5.2.9. *Eyes on the sky: Strategies to mitigate the threats posed by uncrewed aerial systems*

This technical session addressed the threats posed by uncrewed aerial systems (UAS), while also discussing some of the practical and regulatory approaches to countering these threats. Panellists noted that while 'counter drone' systems have become more reliable, there still exist significant barriers to ensuring that the security of nuclear facilities against the use of UAS is consistently maintained. Some of the barriers mentioned during the session included high implementation costs, technical deficiencies in effectively combatting attacks and issues associated with implementing effective regulatory frameworks for governing UAS. One panellist pointed to persistent concerns that UAS technologies are outpacing counter-UAS capabilities, and as such, stakeholders should conduct careful threat assessments, pursue counter-UAS technology integration and push for additional expertise by facility decision makers. Panellists recognized that despite the potential threats, UAS technologies can also serve as valuable tools to enhance the safety and security of nuclear facilities. A key suggestion from the discussion was that the IAEA might hold periodic workshops and develop technical guidance on combatting threats posed by UAS. It was agreed that as the technology rapidly develops, the IAEA can serve as a platform for countries to come together and share national lessons learned in securing critical infrastructure against UAS.

4.5.2.10. *Potential security solutions for traditional and emerging reactor technologies*

This technical session explored key security challenges and potential solutions for various reactor technologies. Topics included addressing potential nuclear security threats when deploying SMRs in some regions; protecting nuclear power plants from social engineering; adapting physical security measures for advanced reactors; potential risks associated with floating nuclear power plants; and integrating proliferation resistance into next generation research reactors. The session highlighted the importance of comprehensive security strategies and international cooperation to mitigate risks associated with both traditional and emerging nuclear technologies. Given the growing interest in SMRs, panellists agreed on the importance of prioritizing this topic at the IAEA and among the broader international community.

4.5.2.11. *Assessing and evaluating physical protection systems in an evolving threat landscape*

During this technical session, panellists presented various means and methods to evaluate the performance and effectiveness of physical protection systems in nuclear facilities. Several different approaches were discussed, including: defining key performance indicators for security processes within an organization; the use of computer programs to evaluate physical protection systems' effectiveness; the development of modelling and simulation capabilities for use in physical protection system exercises, training, and assessments; and implementing performance testing programmes. One presenter detailed possible challenges associated with developing reliable and effective performance testing, due to issues with identifying specialized technical expertise; difficulty in defining clear performance assurance requirements; and extensive time needed to investigate, select, purchase and deploy the equipment necessary for certain performance testing exercises. The concept of 'deterrence' was also discussed, and the panel provided advice around deterrence communication to influence the psychology of threat actors, including insider threats.

4.5.2.12. *Threats or opportunities? Confronting the non-traditional challenges posed by emerging technology*

This technical session addressed various emerging technologies and the issues they pose, including UAS; more affordable, high performance computing and miniaturized electronics; SMRs; and remotely operated weapons systems. Panellists addressed ways in which nuclear power station operators might be enabled through existing legislation to address UAS threats; the new threat profile of SMRs and nuclear fusion reactors; and considerations for defensively using remotely operated weapons without human intervention. While many challenges were noted, it was agreed that opportunities to leverage emerging technology also exist, such as deploying UAS for area surveys or using machine learning to advance real time security assessments of nuclear power facilities.

4.5.2.13. *The threat from within: Addressing and mitigating the insider threat*

During this technical session, panellists discussed newly developed assessment tools and questionnaires to identify potential insider threats both at nuclear facilities and during transport operations. A newly developed empirical model to evaluate the effectiveness of measures to deter threat actors was also presented. Several of the panellists conducted their research within the framework of the IAEA coordinated research project on Preventive and Protective Measures Against Insider Threats at Nuclear Facilities.[6] In addition to presenting research associated with the development of these tools, presenters also discussed various limitations associated with their practical implementation into vetting processes. Further discussion noted that evaluation of personnel should include continuous monitoring, not just initial vetting. A key takeaway was that there may need to be a shift in thinking about personnel, to think of them not as threat vectors, but as a security force multiplier. Moreover, it was noted that there is no 'one size fits all' approach to mitigating the insider threat.

4.5.2.14. *Radiation detection instruments for nuclear security: Strategies for optimizing performance*

This technical session focused on national experiences with deploying detection systems, while also looking to the future to identify areas for improvement. Panellists detailed the work being done today to advance the detection systems of tomorrow. They discussed improving radiation sensors and associated signal processing electronics; developing more powerful computers; aggregating larger datasets; employing more sophisticated online monitoring and analysis software; and developing more comprehensive modelling capabilities. The use of AI and machine learning in detection systems was also addressed, along with the potential to network more

---

[6] For more information on the project, see: www.iaea.org/projects/crp/j02010

instruments. AI may assist with the development of radiation detector algorithms, which can enhance older equipment without necessitating the purchase of new detectors.

4.5.2.15. *Preparing for the future: Lessons learned and developments in nuclear security event response*

This session focused on the national level response to nuclear and radiological incidents, including those incidents caused by malicious acts. The presentations covered cross-cutting approaches to nuclear security and emergency response, along with the importance of preplanned and practised coordination for national response. Panellists presented case studies of incorporating chemical, biological, radiological and nuclear (CBRN) response into federal police protocols; integrating nuclear security expertise into national emergency response organizations; and exercising armed forces to be prepared to respond to nuclear or radiological incidents. An additional case study presented lessons learned from a real world investigation, in which the value of effective coordination and communication between CBRN advisory teams, regulatory bodies and local police forces were highlighted.

4.5.2.16. *Step one: The use of DBT and threat assessment to know the adversary*

Panellists of this technical session represented several countries at varying stages of developing a DBT, while also touching on other threat assessment methodologies. One presenter detailed the steps being taken to revise and peer-review the national DBT for nuclear facilities. Another discussed a project involving multiple national organizations to develop an action plan to create a DBT. Presenters noted that — given new and evolving threat actors, shifts in attacker intentions, and ongoing technological advancement — States should work to establish processes for evaluating and updating their respective DBTs. While a lot of the focus was on means and methods to assess threats to physical protection systems (i.e. through drills, modelling software, and performance testing exercises), one panellist addressed ways to assess computer security risk. The similarities and differences between cyber and physical attacks were noted. Given the strong relationship between operational technology and information technology systems, coupled with the potential for blended attacks from adversaries, it was stressed that physical protection and computer security experts need to work together.

4.5.2.17. *Managing and mitigating counterfeit, fraudulent and suspect items*

This technical session included diverse perspectives on the challenges posed by counterfeit, fraudulent and suspect items (CFSI) in the nuclear security supply chain. Panellists explained that within the nuclear supply chain, CFSI can diminish the integrity of equipment, systems, structures, components or devices that contribute to nuclear safety and nuclear security. Moreover, the inadvertent introduction or malicious insertion of CFSI within the nuclear security supply chain could lead to the occurrence of a nuclear security event. Panellists' research was presented, some of which formed part of an IAEA coordinated research project on the topic. Presenters detailed several mitigation strategies, to include implementing robust supplier vetting and qualification processes, establishing secure authentication mechanisms and implementing rigorous inspection protocols. Research associated with implementing risk assessment models and detection frameworks were also discussed.

4.5.2.18. *Exploring successful implementation strategies for nuclear and other radioactive material transport security*

This technical session explored various implementation strategies for the security of nuclear and other radioactive material during transport. Presenters discussed diverse topics, including using numerical tools and experimental data to enhance future vulnerability assessments, along with a dynamic threat assessment approach incorporating AI. Panellists also presented new methods for monitoring nuclear material in transport and newly developed, advanced e-learning tools to better train and equip practitioners. One presenter discussed the emerging role of uncrewed systems for transport in complex environments, detailing the pros and cons of their employment. Panellists and participants highlighted the importance of sharing best practices for implementing strategies and developing regulatory frameworks for the security of nuclear and other radioactive material in transport, owing to its significance to the global nuclear security regime.

4.5.2.19. Building resilience: Sustaining and enhancing national nuclear security detection architectures

This technical session focused on Member State experiences in developing and sustaining national nuclear security detection architectures (NSDA). Panellists addressed multiple facets of successfully employing an NSDA, including drafting effective standard operating procedures, human resource development, maintenance and performance testing, and assessing new technology for possible adoption into national architectures. A common challenge noted was finding ways to properly communicate threats to decisions makers at all levels in order to gain their support for advancing national NSDA capabilities (i.e. investing in the people, processes and equipment needed to achieve the detection mission). One presenter discussed the creation of a lead national coordination body for nuclear security, which was tasked with managing concepts of operation, standard operating procedures, NSDA evaluations, regional coordination initiatives and other core capabilities. Another discussed the development and deployment of maintenance management software for the detection systems that their State employs.

4.5.2.20. Fortifying the foundations: Implementing physical protection systems and measures

This technical session addressed the physical protection system and measures for material and facilities. Presentations ranged from one State's approach to developing and implementing regulatory frameworks, to newly developed software and evaluation tools to assess physical protection systems during the design process and following implementation. Presenters also discussed more practical topics, including a method to evaluate and optimize surveillance system locations and a novel vehicle inspection system for nuclear facilities. It was noted that the IAEA supports Member States with the evaluation of physical protection systems through various activities, including developing publications, conducting research projects, hosting training activities and conducting IPPAS missions. All those present were encouraged to take advantage of these services.

4.5.2.21. Prepared to respond: National strategies for nuclear security events

During this technical session, panellists shared recent developments, along with existing and emerging challenges, associated with the employment of nuclear security event response plans. Panellists noted that, given the complex international environment and the expansion of other competing serious threats, it is important to ensure that nuclear security threats are not deprioritized. Panellists emphasized that States need to avoid situations where nuclear security is insufficiently integrated in national emergency and crisis response plans. Presentations demonstrated that a strong nuclear security response system involves appropriate levels of involvement from all relevant stakeholders. Collaboration with neighbouring countries can also increase national capacities to respond to nuclear security events. One panellist described the extremely useful employment of regional exercises on nuclear security event response as an efficient tool to train and increase national readiness.

4.6. CAPACITY BUILDING FOR NUCLEAR SECURITY

The nuclear security community faces challenges in sustainably training, developing and retaining the current workforce, while concurrently attracting and retaining young talent, women and diverse personnel. Sessions within this theme examined ways to shape the future with a focus on diversity, equality and inclusiveness.

4.6.1. Plenary panel: Shaping the future — Retaining and developing the nuclear security workforce of tomorrow

This plenary session discussed the challenges the community faces in training and retaining the current workforce; the obstacles to training new employees; and the preparation of the next generation to deal with new and emerging challenges. Panellists discussed various barriers to entry for women and younger professionals into the nuclear security field, as well as strategies to create and maintain inclusive and diverse workforces. Panellists noted the importance of teaching the 'language of nuclear' to those not coming from the nuclear field, in order to effectively bridge the gap between those in technical fields and those in the policy space. Communicating the importance of nuclear security by focusing on real world threats would galvanize the current workforce and increase threat awareness for those both inside and outside of the field. It was suggested that expanding the target recruitment audience to include educating high school students about careers in nuclear security and encouraging non-standard applicants (e.g. mechanical engineers) would be beneficial. In addition, panellists provided an overview of educational programmes, research initiatives and the development of new curricula to prepare the next generation of nuclear security experts. Moreover, panellists highlighted the critical role filled by the International Nuclear Security Education Network (INSEN), which is established by the IAEA, in fostering international cooperation and collaboration to promote sustainable nuclear security education.

4.6.2. Technical sessions

4.6.2.1. Establishing strong nuclear security programmes

Panellists from six different countries and one international organization shared national experiences and discussed various training and capacity building initiatives. Panellists provided updates and progress on their capacity building activities, with each presenter hailing from a State that was at a different stage of developing or pursuing a nuclear power programme. Presenters agreed that with the expansion of the use of nuclear technology for peaceful purposes comes the need for States to ensure their workforce is competent, engaged and vigilant. This applies to those States with established nuclear programmes and for newcomer countries. Depending on where a given country may be in their pursuit of nuclear technologies, it was acknowledged that each country will be faced with unique challenges associated with that specific stage of development. The panel demonstrated that it was important for the international community to collaborate in order that embarking countries can learn from those that have already encountered these challenges. They concluded that establishing robust national education, training and human resource development programmes is and will continue to be an integral part of this process.

4.6.2.2. Equality in action: Strategies for establishing an inclusive workforce

This technical session explored strategies for pursuing and developing an inclusive workforce. The panel shared their diverse perspectives through case studies and national experiences, while also detailing various initiatives implemented at the IAEA. The IAEA was recognized for its ongoing efforts and for the progress it has made in gender parity, with one panellist noting improvements in the diversity of administrative staff and in the inclusion of early career professionals. The session concluded that sharing best practices in gender diversity fosters understanding and advances national and international progress. It was noted that various metrics indicate that the collective efforts of professional women, institutions and the IAEA are driving progress in inclusive workforce development, but continued work is needed.

4.6.2.3. Competence building and human resource management for a robust nuclear security workforce

The panellists in this technical session highlighted key challenges in the field of human resource management for countries seeking to maintain a robust and well-prepared nuclear workforce. Presentations were delivered from representatives of several diverse organizations representing various levels of national nuclear security regimes. These included regulatory bodies, research institutions and law enforcement agencies. Although these organizations and entities have different missions and foci, they each faced several common challenges. One of these challenges is an ageing workforce and the need to retain older professionals in order that institutional knowledge can be passed down to the next generation. Another challenge which was discussed is the retention of trained staff, as personnel are often reassigned or leave the organization for separate ventures. It was noted that such changes create a vacuum, especially when there are limited personnel with such specific knowledge or training on these topics.

4.6.2.4.   Beyond the chalkboard: Modernizing nuclear security training

This technical session introduced a broad range of new approaches and innovative technologies used by Member States to address challenges in training and capacity building. Some of these technologies include remote communication tools, multimedia channels, virtual or augmented reality and simulators. Both Member States and the IAEA have started implementing tools such as and simulations in order to create more realistic training scenarios, create more engaging training exercises, and to yield better results than standard classroom environments. Presenters demonstrated how these tools can be used for both security response and computer security training programmes. In addition, panellists discussed the ways in which various web based platforms enable practitioners to engage in virtual mentoring sessions or distribute video tutorials to wider audiences. In this way, new technology allows for more efficient and sustainable training and capacity building over time. Several panellists discussed the need to continually evaluate the effectiveness of new tools to ensure they are yielding the desired results.

4.6.2.5. Education to excel: Lessons identified from national training centres and institutes

During this technical session, the panellists illustrated global efforts to enhance nuclear security through comprehensive training programmes, international cooperation and the establishment of dedicated facilities to achieve these ends. Panellists highlighted that sustainable training activities are crucial for maintaining high standards, facilitating the secure use of nuclear technologies and contributing to the socio-economic development of their respective countries. The successful implementation of training methodologies, coupled with sustained meaningful international collaboration, has significantly strengthened the knowledge and capabilities of personnel involved in nuclear security worldwide. Several panellists noted that continuous evaluation is needed to assess training programmes and to ensure those managing these programmes are maintaining high standards and constantly improving.

4.6.2.6.   Securing our future: Empowering and enabling the next generation of nuclear security professionals

This technical session discussed the work that several international organizations and networks are doing to develop the next generation of workers. Diversity, equity, inclusion and accessibility were discussed in light of surveys conducted by different national and international organizations. Part of this data indicates that the youth have a positive perception of the nuclear security industry, but also that the nuclear sector faces a 'leaky pipeline' challenge, where women exit the industry at various stages in their career. Several United Nations organizations, alongside the IAEA, have implemented various programmes and initiatives to support the next generation of professionals. Work being done by INSEN was specifically highlighted. It was also suggested that the international community consider drafting language for inclusion in the next nuclear security resolution, which emphasizes the importance of youth engagement.

4.6.2.7.   Shaping the future together: International cooperation in training and capacity building

This session discussed a wide range of past, current and future challenges to capacity building in the field of nuclear security. Several themes were highlighted, including the COVID-19 pandemic, emerging technologies, complexities in international legal frameworks and regional cooperation to address each of these issues. Participants reflected on the ways in which the COVID-19 pandemic drove innovation in developing effective tools for capacity building. For example, online and virtual tools became more widespread, which in turn were used to deliver training on topics such as physical protection and computer security. In addition, panellists noted that other digital approaches, such as 'gamifying' education and training, have also been deployed to engage the next generation. The session stressed the importance of forming communities of practice, particularly in the areas of computer security and AI, which will provide practitioners with networks in which they can cooperate, innovate and develop solutions to unforeseen challenges.

4.6.2.8. Sharing successes and lessons identified from national training initiatives

During this technical session, Member States shared success stories and addressed various issues encountered when implementing national training initiatives. Experts from seven Member States emphasized the critical role of competence building in ensuring strong national nuclear security regimes. This session not only highlighted the IAEA's indispensable role in reinforcing international nuclear security, but it also emphasized the collective responsibility of all States to enhance global nuclear security through their national capacity building programmes. The panellists encouraged all participants to support collaboration between the Agency and Member States to ensure shared approaches to emerging threats and ongoing technological developments.

4.6.2.9. Leveraging interactive methods for enhancing nuclear security capacity building

This technical session explored various case studies and results from national and international training exercises, drills and workshops for nuclear security personnel. Panellists highlighted how interactive training methods encourage personnel to actively participate and to apply their skills in practical scenarios. Such activities provide an opportunity for participants to bridge the gap between theory and real world situations as they can provide organizations with an effective tool to identify outstanding training gaps and areas for improvement. Additionally, although planning exercises often comes with extensive challenges, one presenter discussed the benefits of cross-border training exercises to more easily develop effective plans for national exercise scenarios.

4.6.2.10. From theory to practice: Insights and innovations in nuclear security education

During this session, panellists discussed the evolving nature of nuclear security education programmes. One speaker noted that online and distance learning Masters programmes are effective in attracting younger professionals to the nuclear security field. In addition, online programmes provide a platform on which established nuclear security professionals can advance and expand their expertise. Panellists acknowledged that as professionals progress through their respective careers, they might find it difficult to balance work schedules with in-person classes. Thus, it was argued that there is a need for novel education programmes using non-traditional models, supplemented by micro credentials, professional certificates and accreditation for prior experiential learning. The discussion highlighted that impact assessments and evaluations are an integral part of ensuring the effectiveness and sustainability of both novel and traditional education programmes.

4.7. CROSS-CUTTING NUCLEAR SECURITY TOPICS

The fourth and final theme of the conference included discussions on subjects that spanned multiple dimensions of the nuclear security regime. Some of these topics included the role of the IAEA, international cooperation and collaboration, computer security, nuclear security culture, the role of industry, and the safety–security interface. Such discussions demonstrated the interdisciplinary nature of nuclear security matters, while emphasizing the need for holistic approaches to existing and emerging threats.

4.7.1. Plenary panel: Looking forward — The evolving role of the IAEA

Panellists shared national experiences with using IAEA services and assessed the ways in which the IAEA's role may adapt in the future. They discussed various implications of new technologies and emerging threats for IAEA activities and priorities (e.g. preparing to work with Member States on mitigating the threats associated with the growth of AI). The panel discussed ways in which the IAEA can enhance its role through regional or strategic partnerships, and shared ideas about how the IAEA can become more efficient and effective in providing support to Member States. These included working closely with parliamentarians, further enhancing the interface between nuclear safety and security and simplifying the message about why nuclear security is important for all countries, to include smaller States with no nuclear power programme. It was also suggested that the experiences of the International Civil Aviation Organization and its engagement with stakeholders in industry might be useful for some relevant IAEA activities. No matter the changes that the future might bring, panellists agreed that the IAEA needs to continue to play a central role in advancing global nuclear security and in promoting the peaceful uses of nuclear technology.

4.7.2. Technical sessions

4.7.2.1. *Shared challenges, shared solutions: Regional and international cooperation to enhance nuclear security*

This technical session highlighted the importance and benefits of bilateral, regional and international nuclear security cooperation activities. A recurring theme was the significance of partnerships between national and international organizations to effectively address nuclear security challenges by leveraging each other's expertise and resources. Panellists maintained that capacity building and tailored cross-border training activities are crucial for raising awareness and developing methodologies to enhance national nuclear security capabilities. Regional networks (e.g. the Regional Arms Control Verification and Implementation Assistance Centre (RACVIAC) Nuclear Security Cooperation Initiative, the Forum of Nuclear Regulatory Bodies in Africa, the European Nuclear Security Regulators Association) were also recognized as valuable vehicles for facilitating the sharing of best practices. Such networks can lead to harmonized security standards and collective responses to common threats. Panellists also discussed effective communication and coordination strategies between national stakeholders to ensure that security measures are relevant and impactful, as local engagement aids in the successful implementation of security practices. Panellists agreed that proper documentation and sharing of good practices is essential for continuous improvement and knowledge dissemination across different regions and organizations. The central role of the IAEA in promoting and coordinating global efforts in nuclear security was highlighted, and panellists suggested that the Agency continue to support its Member States, upon request, to further strengthen and sustain their nuclear security infrastructures.

4.7.2.2. *Safety meets security to ensure a safe and secure future*

This technical session covered various perspectives on assessing nuclear safety–security integration and enhancing the nuclear safety–security interface, mainly by drawing on insights from national experiences. Key topics of discussion included safety–security integration assessment tools; security risk assessment methodologies; regulatory approaches for nuclear safety–security interface management; and nuclear safety–security studies in the context of climate change. In addition, one panellist presented a study comparing their country's national standards with recommendations proffered in a published joint report of the IAEA Advisory Group on Nuclear Security (AdSec) and the IAEA International Nuclear Safety Advisory Group (INSAG), entitled A Systems View of Nuclear Security and Nuclear Safety: Identifying Interfaces and Building Synergies. It was suggested that more IAEA Member States conduct such a study to identify potential gaps and discrepancies. The collective insight gained could inform the further development of IAEA safety and security publications, or lead to the creation of joint publications. Another panellist presented the development of their country's national process for the issuance of regulations. In this State, a single authorization is issued for the safety and security of a facility, and joint inspections are carried out involving both safety and security experts.

4.7.2.3. *The IAEA can help you: National case studies on IAEA advisory services and assistance missions*

During this session, presenters shared national experiences of using IAEA advisory services and assistance missions, including IPPAS, the International Nuclear Security Advisory Service (INSServ) and INSSP. Panellists highlighted the benefits, lessons learned, and good practices developed as a result of these and other IAEA advisory or review missions. The panellists emphasized that developing and implementing an INSSP is very beneficial in that it applies a systematic and comprehensive approach to strengthening national nuclear security regimes. Others discussed the steps that their national competent authorities took in the lead up to, and following the completion of, these missions, while noting some of the challenges that were encountered during these processes (e.g. coordination issues between several national competent authorities). Panellists agreed that these services will play a critical role in shaping and reinforcing the global nuclear security regime, and that the IAEA's role in coordinating these activities is invaluable. Panellists encouraged conference participants to consider requesting assistance from the IAEA.

4.7.2.4. *Digital defence: Preparing for and defending against cyber-attacks*

During this technical session, participants emphasized that computer security is an integral component of the safe and secure operation of nuclear facilities. In order to properly prepare practitioners, several panellists discussed national efforts to develop cyber training environments that simulate nuclear facility processes. In addition, one panellist presented a real world example in which an actual computer security incident takes place, along with the steps that the national competent authority took to address this incident. Drawing on this experience, the presenter suggested that the IAEA might consider developing technical guidance for protecting digital assets within a nuclear regulatory authority. Moreover, it was suggested that during future IAEA missions, the Agency might consider including computer security assessments of not only relevant information technology and operational technology infrastructure, but also regulatory information systems. All panel members confirmed their use of various IAEA guidance documents on computer security in nuclear facilities and championed the IAEA's central role in facilitating knowledge exchange amongst computer security experts.

4.7.2.5. *Exploring industry's role in supporting and advancing nuclear security*

This session centred on the role that industry plays in supporting and advancing nuclear security globally. All presenters provided details about the role of industry in their respective countries, while emphasizing that global nuclear security involves collaboration with industry partners. Presenters stressed that in addition to networks within nuclear industry, effective communication and collaboration between industry, government and intergovernmental or international organizations is key to ensuring that nuclear security adapts alongside advancements in technology. Presenters also discussed the importance of engaging and recruiting early career professionals into the field. Participants concurred that the IAEA is one of the many important forums for industry to become more formally involved in global nuclear security, and that the active engagement of industry in the maintenance of the global nuclear security regime should continue to be pursued.

4.7.2.6. *From threats to solutions: Assessing and mitigating cyber-security risk*

Presenters during this technical session highlighted the critical importance of robust and adaptive computer security measures in the nuclear field. Panellists noted that while beneficial, the integration of modern digital technologies and autonomous systems may introduce new challenges that need tailored approaches and continuous vigilance. Several suggestions to mitigate computer security risk were made, including continuous monitoring systems, regularly updating protocols, and establishing computer security operation centres in nuclear facilities. Given the expansion of autonomous systems, one panellist stressed the need to develop clear and consistent frameworks for assessing the computer security risks associated with different levels of autonomy. In addition, there was discussion about developing comprehensive, industry specific guidance and best practices for conducting computer security assessments in industrial irradiator facilities. The session concluded with a call for ongoing collaboration, sharing of best practices and development of comprehensive frameworks to ensure computer security risk is effectively managed.

4.7.2.7. *Security — It's what we do: Developing and maintaining a robust nuclear security culture*

This technical session covered various perspectives on assessing and enhancing nuclear security culture. Panellists discussed insights from national approaches and proposed models for self-assessment strategies. Key themes included the importance of continuous training, the role of raising public awareness and the development of tailored initiatives to foster a robust security culture. Presentations highlighted the need for analysis of incidents and their causes, enhanced international collaboration and cooperation with the public, and innovative training methods utilizing tabletop exercises and 'gamified' scenarios. Panellists stressed that nuclear security culture plays an important role in ensuring that individuals, organizations and institutions remain vigilant and that measures are sustained to prevent malicious acts. One panellist noted that the IAEA stands ready to support any country looking to reinforce its nuclear security culture.

4.7.2.8. Exploring the role of civil society in shaping the future of nuclear security together

In this technical session, panellists discussed the role of think tanks, universities and civil society in enhancing nuclear security regionally and globally. Discussions centred around leveraging epistemic communities for expert knowledge generation and policy recommendations; addressing challenges, such as funding sustainability and policy impact uncertainty; and fostering inclusive cooperation and consistent leadership to strengthen nuclear security frameworks and norms. Some suggestions proffered by the panellists included prioritizing nuclear security during heightened risks, enhancing security culture, and promoting an inclusive narrative to shape a more secure future for all. Participants concurred that civil society can bring added value to nuclear security through collaboration, education, communication and effective messaging.

4.7.2.9. Let's talk about nuclear security: Considerations for public communication

This session underscored the importance of effective and adaptive communication strategies in nuclear security, especially given the expansion of nuclear technologies to meet the United Nations Sustainable Development Goals. Panellists discussed the need for ongoing and continuous public engagement, and the importance of addressing misinformation to maintain public confidence and safety. One speaker discussed ongoing research associated with the use of humour as a strategy to reduce fears and improve message delivery. It was noted that communications on nuclear security need to strike a delicate balance owing to the need for transparency and given the potentially confidential information involved. Overall, panellists emphasized the advantages of using clear, culturally sensitive and empathetic communication strategies in nuclear security, highlighting the importance of proactive engagement, accurate information dissemination and technical expertise to maintain public trust.

4.7.2.10. Adapting to adversity: Ensuring nuclear security in complex and challenging environments

This technical session explored lessons learned from past challenges in order to enhance contemporary nuclear security systems. One panellist detailed the extensive work that had been done to develop nuclear security training for personnel in conflict zones. Other panellists discussed some of the strategies that their respective States employed to maintain nuclear security throughout the COVID-19 pandemic. Common experiences underlined the value of new digital technologies to sustain regulatory functions in periods when in-person interaction is very difficult or impossible. At the same time, the common view was that in-person interaction was vital to maintaining effectiveness over time and cannot be fully replaced by digital technology or virtual meetings.

4.7.2.11. Developing approaches to risk and threat assessments for robust nuclear security

During this session, examples of various approaches to risk and threat assessments for several different components of the nuclear security regime were presented (e.g. transport security, computer security). Panellists discussed the use of innovative technologies and advanced security modelling as a means to assess the efficacy of physical protection system designs. Other panellists discussed the development of frameworks for the conduct of risk assessments at production sites, along with computer security assessments at other radioactive material facilities. Each of these discussions highlighted both national and international efforts to improve approaches to assessing risk, in order that decision makers at all levels are properly equipped to make informed decisions about the implementation of nuclear security measures.

4.8. CLOSING PLENARY SESSION

Following a brief introduction by Ms Lydie Evrard, Deputy Director General and Head of the Department of Nuclear Safety and Security, the ICONS 2024 Nuclear Security Delegation for the Future delivered a statement that they had cooperatively drafted in the months leading up to the conference. Their statement highlighted the importance of providing various pathways to young professionals to join the nuclear security sector and of providing equitable access to workshops and training. They also stressed the importance of intergenerational discussions, especially on topics related to emerging technologies, as their generation will inherit the consequences of decisions made about their use.[7]

HE Mr Ian Biggs, Ambassador and Permanent Representative of Australia, and HE Mr Mukhtar Tileuberdi, Ambassador and Permanent Representative of Kazakhstan, then delivered remarks as representatives of the Co-Presidents of the conference. Director General Rafael Grossi was then invited to deliver the closing remarks for ICONS 2024, following which he declared the conference officially closed.[8]

---

[7] The statement of the Nuclear Security Delegation for the Future is presented in Section 6 of this publication.
[8] The closing statements are presented in Section 3 of this publication.

5. STATEMENT BY THE CO-PRESIDENTS

(1)   We the Co-Presidents of the International Atomic Energy Agency (IAEA) International Conference on Nuclear Security: Shaping the Future, reiterate our commitment to sustain and strengthen effective and comprehensive nuclear security of all nuclear and other radioactive material and facilities.

(2)   We reassert that the responsibility for nuclear security within a State rests entirely with that State in accordance with its respective national and international obligations, mindful of the sovereign rights of every Member State.

(3)   We reaffirm the common goals of nuclear non-proliferation, nuclear disarmament and peaceful uses of nuclear energy, recognize that nuclear security contributes to international peace and security, and stress that progress in nuclear disarmament is critically needed and will continue to be addressed in all relevant fora, consistent with the relevant obligations and commitments of Member States.

(4)   We acknowledge that nuclear security measures may enhance public confidence in the peaceful use of nuclear applications. We also acknowledge that those applications contribute to Member States' sustainable development, and we should ensure that measures to strengthen nuclear security do not hamper international cooperation in the field of the peaceful uses of nuclear applications.

(5)   We remain concerned about existing and emerging nuclear security risks and threats and commit to addressing them. We recognize that emerging and innovative technologies, inter alia, artificial intelligence, present potential challenges and benefits. In this regard, we further recognize the importance of international cooperation in support of national nuclear security regimes, to help us maximize benefits while addressing challenges.

(6)   We support the work of the IAEA in assisting Member States, upon request, in establishing and improving effective and sustainable national nuclear security regimes, including through guidance development, advisory services, and capacity building, and accordingly its central role in facilitating and coordinating international cooperation to strengthen nuclear security, as well as its role in facilitating, as appropriate, regional activities.

(7)   We recognize physical protection as a key element in nuclear security, and support the further development of the IAEA's assistance in the relevant areas of importance to Member States to include prevention, deterrence, detection, access delay and response.

(8)   We emphasize that any attacks or threats of attacks against nuclear facilities devoted to peaceful purposes may compromise nuclear security and recall our commitments in this regard. We note General Conference resolutions GC(XXIX)/RES/444 and GC(XXXIV)/RES/533 and the 2009 General Conference unanimous decision GC(53)/DEC/13.

(9)   We recognize the need to ensure resilience in national nuclear security regimes and emergency preparedness in all circumstances, including extraordinary circumstances. We note the IAEA Director General's "seven indispensable pillars for ensuring nuclear safety and security during an armed conflict".

(10)   We encourage Member States to implement threat mitigation and risk reduction measures that contribute to improving nuclear security including, but not limited to, ensuring the protection of nuclear and other radioactive materials and facilities in accordance with national legislation.

(11)   We call upon all Member States possessing HEU and separated plutonium in any application, which require special precautions to ensure their nuclear security, to make sure they are appropriately secured and accounted for, by and in the relevant State, and we encourage Member States, on a voluntary basis, to further minimize HEU in civilian stocks, when technically and economically feasible.

(12)   We emphasize the importance of nuclear security considerations in the context of the growing interest in the development and deployment of advanced nuclear technologies and reactors, including small and modular reactors (SMRs), in accordance with the respective obligations of Member States.

(13)   We recognize the threats to computer security and from cyber-attacks at nuclear facilities, as well as

their associated activities including the production, use, storage and transport of nuclear and radioactive materials, and highlight the need by Member States to continue addressing computer security risks when strengthening the protection of sensitive information and computer based systems, and encourage the IAEA to continue to foster international cooperation and to assist Member States, upon request, in this regard.

(14) We reaffirm the importance of continuing to promote the universalization and implementation by its States Parties of the Convention on the Physical Protection of Nuclear Material (CPPNM) and its Amendment. We note the convening of the 2022 Conference of the Parties to the Amendment to the CPPNM which concluded that the Convention as amended was adequate and look forward to the second Conference of the Parties and the review of the implementation and adequacy of the Convention. We also reaffirm the importance of other relevant international legal instruments, such as the International Convention for the Suppression of Acts of Nuclear Terrorism (ICSANT).

(15) We commit to maintaining effective security of radioactive sources throughout their life cycle, consistent with the objectives of the Code of Conduct on the Safety and Security of Radioactive Sources and its supplementary guidance documents.

(16) We emphasize the importance of security in the transport of nuclear and other radioactive material and, in recognition of the continuous increase in the amount of such material, stress the need for Member States to take effective measures, consistent with their international and domestic obligations.

(17) We encourage the IAEA to continue facilitating, in close cooperation with Member States, a coordination process to address the interface between nuclear security and nuclear safety, as appropriate.

(18) We reiterate our commitment to combatting illicit trafficking of nuclear and other radioactive material and to ensure that the material cannot be used by non-State actors for malicious purposes and encourage Member States to continue sharing relevant information, on a voluntary basis, including through relevant channels. The States providing notifications to databases are responsible for accuracy, objectivity and purely technical character of this information.

(19) We support the IAEA's and Member States' efforts to strengthen nuclear security culture in the framework of organizational culture in a balanced and risk-informed manner, and also insider threat prevention and mitigation, in particular through providing education and training opportunities, and note the contribution of other relevant institutional entities, such as regulators and industry, in this regard.

(20) We welcome the opening of the IAEA Nuclear Security Training and Demonstration Centre (NSTDC), thereby complementing the existing work of Member States' nuclear security support centres, Centres of Excellence, and IAEA Collaborating Centres, and supporting the IAEA's efforts in capacity building to strengthen national nuclear security regimes, thus emphasizing the importance that Member States support the centre.

(21) We encourage the IAEA Secretariat and Member States to continue their efforts in education and development of current and future generations of nuclear security professionals.

(22) We encourage Member States to use and contribute to the IAEA's nuclear security review missions and advisory services, on a voluntary basis.

(23) We call upon Member States to support and contribute, as appropriate, to the IAEA's nuclear security activities by providing experts and sharing national expertise, best practices, lessons learned, as well as highlighting recent successes, with due regard to the protection of sensitive and confidential information.

(24) We recognize the Nuclear Security Fund as an important instrument for the Agency's activities in the field of nuclear security. We will continue to provide, on a voluntary basis, funds to the Nuclear Security Fund, as well as technical and human resources, as appropriate for the IAEA to implement its work in nuclear security and to provide, upon request, the support needed by Member States.

(25) We commit to promote equitable geographical distribution and gender equality in the context of IAEA's nuclear security activities, and encourage Member States to establish an inclusive workforce within their national nuclear security regimes, including ensuring equal access to education and training.

(26) We call upon the IAEA Secretariat and Member States to take this Statement into account in the consultation process between the Secretariat and the Member States during the development of the IAEA's 2026–2029 Nuclear Security Plan, while also considering the proceedings of this conference, as appropriate.

(27) We call upon the IAEA to continue to improve communication with Member States about its nuclear security activities and to facilitate the exchange of technical and scientific information on nuclear and radiation technology options in the field of nuclear security.

(28) We call upon the IAEA to continue to organize international conferences on Nuclear Security every four years and encourage all Member States to participate at a Ministerial level.

6. STATEMENT BY THE NUCLEAR SECURITY DELEGATION FOR THE FUTURE

The Nuclear Security Delegation for the Future is a diverse team of 24 young nuclear security professionals from 19 countries, identified by the International Atomic Energy Agency (IAEA) through a competitive application process as part of an inaugural initiative organized for the 4th International Conference on Nuclear Security (ICONS) 2024: Shaping the Future.

We, the Nuclear Security Delegation for the Future, acknowledge that the ICONS 2024 theme of 'Shaping the Future' provides an opportunity for intergenerational dialogue and cooperation in order to achieve peace and security worldwide. We believe that including young, early career professionals into the heart of these discussions is essential.

We collaborated closely for three months, drawing on our individual academic and professional backgrounds to discuss the future of nuclear security, to harmonize our perspectives and to ultimately present our deliberations at ICONS 2024. Through this joint effort, our Delegation tangibly demonstrates cooperative global leadership among young nuclear security professionals while actively contributing to the international discussion on nuclear security.

The Nuclear Security Delegation for the Future statement addresses four sub-areas: global communication; emerging technologies and artificial intelligence; capacity building; and how to make nuclear security inclusive for all.

Global communication for nuclear security

(1) We urge the international community to endeavour to amend global perceptions on nuclear technologies through education and knowledge sharing, both about these technologies' benefits and about the measures that are taken to manage the risks.

(2) We advocate for the recognition of nuclear safety, security and safeguards as unavoidably connected and as critical for the protection of the people, environment and society.

(3) We call upon Member States to facilitate transparency in the field, and to continue public engagement through effective communication strategies and outreach programmes in order to facilitate the understanding of nuclear risks and related mitigation strategies implemented by the global community.

(4) We reassert the need for international cooperation across borders, which encompasses sharing knowledge, best practices and lessons learned from incidents, as well as taking collective action to enhance nuclear security, thereby ensuring a safer and more secure world for generations to come.

(5) We recognize the importance of academic research, collaborations and training to advance nuclear security. These contributions drive innovation, foster interdisciplinary approaches, develop cutting-edge technologies and prepare the next generation of professionals to address emerging threats. These efforts are essential for ensuring the continued protection of nuclear material and facilities.

(6) We stress the importance of international treaties, agreements and regulatory frameworks in promoting global nuclear security standards and best practices. We also acknowledge the persistent challenges posed by the evolving nature of international relations. We call upon Member States to ensure the peaceful use of nuclear technology through the strengthening of international legal frameworks in order to meet the shifting needs of the global community.

(7) We uphold the value of intergenerational discussion and call for the next generation of nuclear professionals to become stakeholders in nuclear security policy and decision making discussions.

Emerging technologies and artificial intelligence for nuclear security

(8) The development of novel designs of small modular reactors (SMRs) and advanced modular reactors (AMRs) raises important security considerations in many areas, including facility operations, transportation and waste storage. Security requirements and recommendations must evolve alongside technological advancements. Addressing these evolving challenges presents opportunities for young professionals to influence the future of nuclear security.

(9) The pivotal role that SMRs and AMRs will play in shaping our sustainable energy future will be driven by young nuclear security professionals whose essential expertise will ensure these transformative technologies can be deployed safely and securely.

(10) We firmly believe in empowering young professionals to actively participate in the forefront of nuclear security research and to engage in conferences, seminars and webinars dedicated to nuclear security, SMRs and advanced reactors in order to stay apprised of latest developments in the field.

(11) We highlight the importance of international cooperation. We, the next generation, are eager to foster collaboration and knowledge sharing on SMR and AMR technologies, safety standards, nuclear security guidance and non-proliferation efforts globally.

(12) We similarly recognize the dual nature of artificial intelligence (AI). AI could greatly benefit the field of nuclear security if designed, developed and deployed securely; however, the speed and scale at which AI is developing poses a potential grave threat which needs to be addressed.

(13) We call upon Member States to cooperate to mitigate this potential threat, and to jointly pledge to utilize technologies such as AI solely for peaceful purposes. We advocate for the promotion of the peaceful uses of emerging technology through the strengthening of international legal frameworks, and urge States to never use such technology to target critical infrastructure.

(14) We urge the international nuclear security community to continue to invest in and maintain analogue and legacy security systems in order to ensure continued operability, sustained security and the diversity of operational technology.

(15) We ask that the next generation of nuclear security professionals be involved in decision making on the use of emerging technologies. We, the next generation, will inherit the consequences of their development. As such, we must be given the power and resources to guide and influence this process.

Capacity building for nuclear security

(16) We consider attracting young talent to nuclear security to be a priority.

(17) We believe that through targeted outreach and engagement strategies, young individuals at various educational levels should be presented with opportunities to join the nuclear security sector and be introduced to pathways into this field.

(18) We recognize that recruiting, retaining and developing talented individuals, particularly the younger generations entering the nuclear field, can be a challenging endeavour. We call for the IAEA and Member States to explore how to tailor engagement strategies to recruit the next generation into nuclear security. This includes utilizing educational games, social media outreach and other opportunities beyond conventional governmental frameworks.

(19) We need to ensure that young talent has the support and resources to forge pathways into leadership roles. As the workforce is maturing, youth voices will become the decision making bodies of the future, and equipping them with the tools to effectively lead is essential.

(20) We ask that the efforts which are put into training and educating these individuals reflect the importance of their presence in the nuclear security field. Technical exchanges, cross-sector collaborations and other developmental opportunities need to be plentiful and made available to everyone entering the field. Younger generations should have equitable access to nuclear security training and workshops.

(21) We encourage interdisciplinary collaboration by integrating young professionals from adjacent fields,

including IT and engineering, into roles associated with nuclear security. This approach not only broadens perspectives but also fosters sustainable solutions to nuclear security challenges. The future of nuclear security requires a variety of disciplines working in parallel and across safety–security interfaces in order to create a unified approach. Such collaborative efforts will enable technological applications that were previously thought not possible.

Nuclear security for all

(22) We, the Delegation for the Future, highlight that nuclear security should be a priority national concern due to its global implications. Existing and newly developed national standards and regulations should be oriented towards upholding international legal instruments, to include the Convention on the Physical Protection of Nuclear Material (CPPNM) and its Amendment. If we allow discrepancies in implementation to increase, there exists the likelihood for potential threats to endanger current and future generations.

(23) We, as a transnational delegation, are driven by a vision that transcends borders and which emphasizes cooperative approaches that bolster nuclear security for all. We note the importance of building upon the foundations of traditional approaches to nuclear security while also adapting to confront new and emerging challenges.

(24) We encourage a holistic approach to nuclear security, which above all else empowers all communities, thereby guaranteeing their rightful inclusion in the nuclear security dialogue.

(25) We plead with Member States to deepen cooperation to ensure the continued safe and secure operations of nuclear power plants, even in extraordinary circumstances. It is equally important to learn from these consequential incidents and establish comprehensive nuclear security guidelines with the appropriate oversight and direction to continue learning from these experiences.

(26) We urge Member States to increase international nuclear security personnel training. These collaborative efforts will ensure that technical staff (including first responders and security forces) are prepared with pre-established cooperation frameworks to counter myriad scenarios and threats to critical nuclear infrastructure. Collaboration fosters a sense of shared responsibility and increases knowledge sharing, thereby enhancing the overall security landscape.

(27) Increased diversity is an essential part of our future strategy as it leads to better decisions and more positive outcomes. The global community must continue on its upward trajectory of increasing diversity in the nuclear security sector in pursuit of a world in which all can benefit from nuclear technologies, and in which no one is left behind.

We implore you to consider our appeals as it is our future at risk

We are thankful to the IAEA for launching this exceptional opportunity for young professionals to gain real world experience in leadership, diplomacy and international nuclear security. We extend our heartfelt gratitude to the staff of the Division of Nuclear Security and the Conference Secretariat: Mr Bryan Denehy, Mr Christian Deura, Ms Sara Mroz and Mr Sanjai Padmanabhan for their exceptional support through the Nuclear Security Delegation for the Future deliberations and development of this Statement.

Contributors

Abubakar Sadiq Aliyu	Emanuel Lukawiecki	Mariana Sorroza
Zhanna Asulian	Diana Mafie	Win Thu Zar
Camila Boix	Wendolyn Martinez	Megan Tougas-Cooke
Musa Carew	Wilfred Massiata	Oceane Van Geluwe
Amanda Carvalho	Tibyan Mustafa	Cosmin Vasiliu
Megan Clarke	Mustafa Ozbek	Alex Vipond
Megan Fearon	Yerdaulet Rakhmatulla	Huirong Wang
Sheena Lim	Facundo Saponara	Nicole White

7. CONTENTS OF THE SUPPLEMENTARY FILES

The supplementary files for this publication are available at: www.iaea.org/publications. The content of the supplementary files is organized as follows:

— Conference programme;
— List of delegations' participants;
— Photographs featured in the photography competition entitled 'Nuclear Security Through the Lens'

ANNEX I.

CONFERENCE STATISTICAL DATA

Organized by	IAEA
Held in cooperation with	Division of Nuclear Security Department of Nuclear Safety and Security
Location	Vienna International Centre, M Building
Scientific Secretary	Sara Mroz, NSNS
Scientific support	Bryan Denehy, NSNS Christian Deura, NSNS
Conference coordination	Martina Neuhold, CSS-MTCD Sanjai Padmanabhan, CSS-MTCD
Conference support	Erik Paniagua-Miranda, CSS-MTCD Lara Popescu, CSS-MTCD
Conference web page	www.iaea.org/events/icons2024

I–1. MINISTERIAL SEGMENT

Registration and attendance

Total no. of official participants	1258
No. of Member States	131
No. of permanent observers	1
No. of organizations	7
Thereof female	473 (38%)
Thereof male	785 (62%)
In-person attendance	1207
Virtual attendance	51

Detailed registration information

> **1209** participants from **131** Member States

Afghanistan	2	Greece	3	Pakistan	11
Albania	3	Guatemala	4	Palau	2
Algeria	8	Holy See	7	Panama	3
Angola	3	Honduras	4	Paraguay	7
Argentina	9	Hungary	9	Peru	2
Armenia	12	Iceland	2	Philippines	7
Australia	30	India	3	Poland	6
Austria	11	Indonesia	7	Portugal	3
Azerbaijan	12	Iran, Islamic Republic of	9	Qatar	8
Bahamas	2	Iraq	8	Republic of Moldova	2
Bahrain	3	Ireland	5	Romania	6

Bangladesh	3	Israel	16	Russian Federation	33
Belarus	4	Italy	6	Saint Lucia	4
Belgium	9	Japan	16	Saudi Arabia	8
Bosnia and Herzegovina	3	Jordan	11	Senegal	6
Botswana	7	Kazakhstan	21	Singapore	6
Brazil	14	Kenya	6	Slovakia	9
Brunei Darussalam	1	Korea, Republic of	47	Slovenia	4
Bulgaria	6	Kuwait	4	South Africa	10
Burkina Faso	5	Lao People's Democratic Republic	4	Spain	15
Cambodia	3	Latvia	3	Sri Lanka	3
Canada	13	Lebanon	6	Sudan	4
Chile	5	Lesotho	5	Sweden	17
China	12	Liberia	1	Switzerland	13
Colombia	4	Libya	5	Syrian Arab Republic	5
Costa Rica	4	Liechtenstein	2	Tajikistan	7
Cote d'Ivoire	1	Lithuania	6	Thailand	7
Croatia	7	Luxembourg	4	Togo	3
Cuba	6	Malaysia	14	Tunisia	3
Cyprus	4	Malta	4	Turkmenistan	6
Czech Republic	14	Mexico	5	Türkiye	7
Denmark	4	Mongolia	2	Uganda	2
Dominican Republic	3	Morocco	11	Ukraine	8
Ecuador	2	Myanmar	5	United Arab Emirates	31
Egypt	5	Namibia	5	United Kingdom	41
El Salvador	5	Nepal	4	United Republic of Tanzania	2
Estonia	9	Netherlands, Kingdom of the	14	United States of America	171
Eswatini	1	New Zealand	3	Uruguay	3
Ethiopia	5	Nicaragua	1	Uzbekistan	3
Finland	15	Niger	1	Venezuela, Bolivarian Republic of	4
France	50	Nigeria	4	Viet Nam	4
Georgia	4	North Macedonia	3	Yemen	3
Germany	31	Norway	12	Zimbabwe	16
Ghana	18	Oman	10		

> ➢ **3** participants from **1** permanent observer

State of Palestine	3

> ➢ **46** participants from **7** organizations

European Union	14
Inter-Parliamentary Union (IPU)	8
International Criminal Police Organization (INTERPOL)	1
League of Arab States (LAS)	3
Security Council Committee established pursuant to resolution 1540 (2004) (1540 Committee)	1
United Nations Interregional Crime and Justice Research Institute (UNICRI)	1
World Institute for Nuclear Security (WINS)	18

Special event — Beyond borders

Total no. of official participants	184
No. of Member States	64
No. of organizations	3

Special event — Interactive session

Total no. of official participants	138
No. of Member States	61
No. of organizations	3

I–2. SCIENTIFIC SEGMENT

Total no. of official participants (including invited persons)	818
Total no. of invited persons	47
No. of Member States	111
No. of permanent observers	0
No. of organizations	15
Thereof female	247 (30%)
Thereof male	571 (70%)
In-person attendance	619
Virtual attendance	199

Number of participants by Member State

> **713** participants from **104** Member States (without invited persons)

Albania	2	Greece	5	Paraguay	1
Algeria	1	Hungary	7	Philippines	1
Argentina	5	India	9	Poland	2
Armenia	1	Indonesia	59	Qatar	1
Australia	3	Iran, Islamic Republic of	2	Republic of Moldova	1
Austria	6	Iraq	4	Romania	7
Azerbaijan	2	Italy	10	Russian Federation	7
Bangladesh	3	Jamaica	1	Rwanda	2
Belarus	1	Japan	14	Saint Kitts and Nevis	3
Belgium	3	Jordan	6	Saint Lucia	1
Benin	2	Kenya	18	Saudi Arabia	5
Brazil	16	Korea, Republic of	10	Serbia	2
Burkina Faso	1	Lao People's Democratic Republic	1	Singapore	3
Burundi	2	Latvia	2	Slovakia	1
Cambodia	2	Lebanon	1	Slovenia	6
Cameroon	7	Lesotho	2	South Africa	2

Country		Country		Country	
Canada	30	Libya	2	Sri Lanka	6
Central African Republic	2	Madagascar	1	Sudan	5
Chad	1	Malawi	3	Sweden	1
China	54	Malaysia	10	Tajikistan	7
Colombia	5	Mali	2	Thailand	2
Comoros	1	Mauritania	2	Togo	2
Congo	4	Mexico	1	Tunisia	2
Cuba	4	Montenegro	4	Türkiye	8
Czech Republic	1	Morocco	9	Uganda	11
Democratic Republic of the Congo	4	Mozambique	3	United Arab Emirates	3
Dominican Republic	2	Myanmar	1	United Kingdom	60
Egypt	16	Namibia	1	United Republic of Tanzania	5
Estonia	1	Netherlands, Kingdom of the	10	United States of America	27
Ethiopia	9	Nicaragua	1	Uzbekistan	2
France	32	Niger	1	Viet Nam	1
Gabon	2	Nigeria	22	Yemen	1
Georgia	4	North Macedonia	2	Zambia	2
Germany	16	Norway	8	Zimbabwe	6
Ghana	23	Pakistan	15		

➢ **58 participants from 15 organizations (without invited persons)**

Organization		Organization	
Institute of Nuclear Materials Management (INMM)	1	United Nations Interregional Crime and Justice Research Institute (UNICRI)	3
International Criminal Police Organization (INTERPOL)	9	United Nations Office of Counter-Terrorism	1
International Maritime Organization (IMO)	1	United Nations Office for Disarmament Affairs (UNODA)	9
James Martin Center for Nonproliferation Studies (CNS)	3	United Nations Office on Drugs and Crime (UNODC)	4
Nuclear Threat Initiative (NTI)	8	Vienna Center for Disarmament and Non-Proliferation (VCDNP)	12
OECD Nuclear Energy Agency (NEA)	2	Women in Nuclear	1
Organization of American States (OAS)	2	World Association of Nuclear Operators (WANO)	1
Stockholm International Peace Research Institute	1		

ANNEX II.

LIST OF TECHNICAL AND SCIENTIFIC PAPERS

The technical and scientific papers[9] from the International Conference on Nuclear Security: Shaping the Future (ICONS 2024) are available on the IAEA website.

Abstract ID	Title of paper	Contributor(s) and affiliation(s)
6	Nuclear security implications of counterfeit, fraudulent and suspect items entering the nuclear supply chain	C. HOBBS Z. NASER D. B. SALISBURY S. TZINIERIS *King's College London* United Kingdom
7	The performance of the Brazilian armed forces in the planning and preparation of the medical response to radiological and nuclear accidents	M. THADEU CARTAXO DA COSTA *Ministry of Defense* Brazil
8	Enhancing and maintaining of spectrometry expertise through spectral flavor of the month: CSN's experience	M. KACI *Nuclear Security Center* Algeria
9	Off-site nuclear security responders: Workforce management and capabilities building	C. ROMÃO *Institutional Security Cabinet* Brazil
10	Training in support of enhanced nuclear security of radioactive materials at CEADEN	D. SOGUERO *Centre of Technological Applications and Nuclear Development (CEADEN)* Cuba P.L. JORGE LUIS *Office for Environmental Regulation and Safety* Cuba
14	Basis for graded approach for security requirements	D.A. BOKOV A.A. SIMONOV A.A. EGOROV *Federal Environmental, Industrial and Nuclear Supervision Service (Rostechnadzor)* Russian Federation
15	Current status of radiation detection instruments in Libya: Challenges and prospects	K. ELMASRI *University of Tripoli* Libya
16	A software development for physical protection system design and analysis for nuclear security	S. ŞENTÜRK LÜLE M. AKDEMİR *Istanbul Technical University, Energy Institute* Türkiye

---

[9] The technical and scientific papers are available for consultation by registered participants of the conference either on the conference application or on the Nuclear Security Information Portal (NUSEC).

Abstract ID	Title of paper	Contributor(s) and affiliation(s)
17	An overview of isotope identification algorithms used in radionuclide identification devices based on medium energy resolution detectors: Detection of radioactive/nuclear material and radionuclide categorization in nuclear security	H. YÜCEL *Ankara University* Türkiye
20	Small modular reactor (SMR) deployment in Sub-Saharan Africa (SSA): Addressing nuclear security threats	S. ADU C. MARIANNO *Texas A&M University* United States of America
24	The role of think tanks and universities in enhancing nuclear security in southeast Asia	J.C. TRAJANO M. CABALLERO-ANTHONY *Nanyang Technological University* Singapore
26	DBT approach in the Netherlands	S. VLEUGELS *Authority for Nuclear Safety and Radiation Protection* Kingdom of the Netherlands
27	Action plan for the implementation of the DBT process in Brazil	R.W.D. GARCÊZ J.S. MONTEIRO FILHO L.F.B. TORRES A.R. DE LIMA *National Nuclear Energy Commission* Brazil R.L.A. TAVARES *Institutional Security Office* Brazil
28	International Physical Protection Advisory Service (IPPAS) mission: Nigerian experience	N.A. BELLO E.E. OFOEGBU *Nigerian Nuclear Regulatory Authority* Nigeria
31	Student engagement program in nuclear security at ISCN/JAEA	NAOKO INOUE MEGUMI SEKINE NAOKO NORO YOSUKE NAOI RIE SUGIYAMA TAKAHARU TATSUNO *Japan Atomic Energy Agency* Japan
32	Legislative and regulatory framework for nuclear security: A case study of Tanzania	J.M. MWIMANZI *Tanzania Atomic Energy Commission* United Republic of Tanzania
38	Research for better vulnerability assessment regarding acts of sabotage: Research on nuclear security with regard to nuclear transports	C.O. GERBER A. BECKER M. PELZER *Installation and Reactor Safety Company (GRS)* Germany

Abstract ID	Title of paper	Contributor(s) and affiliation(s)
39	Effective and efficient radiological security capacity building support for Asian region: US–Japan partnership	NAOKO NORO MEGUMI SEKINE NAOKO INOUE *Japan Atomic Energy Agency* Japan B. BUDDEMEIER *Lawrence Livermore National Laboratory* United States of America
42	Approaching risk analysis through integration of nuclear safety and security	T. THOMAS J. HARRIS *Purdue University* United States of America
44	A methodology to assess insider threat risk at nuclear facilities	R. AGALGA D.N. ADJEI P.A. AMOAH D.F. CHARLES I.S. SURAJ *Ghana Atomic Energy Commission* Ghana
45	Capacity and competence building in nuclear security in developing countries	A.I. AMASI *The Nelson Mandela African Institution of Science and Technology* United Republic of Tanzania
46	Holistic approach toward sustainable and effective nuclear security regime in Lebanon	M. ROUMIE B. NSOULI *Lebanese Atomic Energy Commission* Lebanon
48	Recent nuclear forensics technical capability building efforts by Integrated Support Center for Nuclear Nonproliferation and Nuclear Security at Japan Atomic Energy Agency	YOSHIKI KIMURA YOSHIKI MATSUI TOMOKI YAMAGUCHI *Japan Atomic Energy Agency* Japan
50	Development of video tutorials to promote use of information management tools provided for the NSSC network	MEGUMI SEKINE-ABE JUNKO ASHIMA NAOKO NORO NAOKO INOUE *Japan Atomic Energy Agency* Japan YURA SHIN *International Atomic Energy Agency* IN YOUNG SUH *Oak Ridge National Laboratory* United States of America R. MELLOUKI *National Centre for Nuclear Energy Sciences and Technology* Morocco

Abstract ID	Title of paper	Contributor(s) and affiliation(s)
51	Development of ISCN exercise field	TAKUYA KOBAYASHI NAOKO INOUE HIDETOSHI SUKEGAWA YASUSHIRO HOSOKAWA NOBUHIKO HASEGAWA YUKO KOMORI YO NAKAMURA MIKU MIZUEDANI MEGUMI SEKINE NAOKO NORO *Japan Atomic Energy Agency* Japan
52	Actions of the regulatory body on the enhancement of the sustainable physical protection regime and nuclear security culture through the interaction with operators	M.V. IVANOV *Federal Environmental, Industrial and Nuclear Supervision Service (Rostechnadzor)* Russian Federation
55	Nuclear security risk assessment of counterfeit, fraudulent, and suspect items (CFSI) infiltration within supply chain: Bangladesh perspective	D. HOSSAIN *Bangladesh Atomic Energy Commission* Bangladesh
58	Advancing international standards for social media vetting in nuclear security: A proposal for mitigating insider threats	J. LANDERS *Oak Ridge National Laboratory* United States of America J. BAWEJA *Pacific Northwest National Laboratory* United States of America J. KINNEY M. FIALKOFF *Oak Ridge National Laboratory* United States of America
62	Security risk mitigation of counterfeiting and fraudulent items within the nuclear supply chain	A. ARAFA *Egyptian Atomic Energy Authority* Egypt
63	Radioactive sources security regulation developed and implemented in Cameroon: Regulation development and implementation	J.F. BEYALA ATEBA A. SIMO *National Radiation Protection Agency* Cameroon
64	Fully autonomous unmanned aircraft systems and the consequences for national security	B. MAXWELL *United States Department of Energy* United States of America
65	Cooperation among United States federal and state agencies to protect radioactive materials through the task force on radiation source protection and security	G. SMITH *United States Nuclear Regulatory Commission* United States of America
66	Enhancing nuclear security through continuous education and international collaboration	E. STRUHAROVÁ D. MACÁK *Presidium of the Police Force* Slovakia

Abstract ID	Title of paper	Contributor(s) and affiliation(s)
72	Nuclear security defense in depth and graded approach during the past commonwealth heads of government meeting 2022 in Rwanda	R. BANA E. ABENANYE *Rwanda Utilities Regulatory Authority* Rwanda
76	The methods of physical protection system effectiveness evaluation and application of these methods in the process of the physical protection system development and operation	A.V. IZMAYLOV *JSC Federal Center for Science and High Technology Special Scientific and Production Enterprise (Eleron)* Russian Federation
79	Impact of international cooperation assistance in enhancing national regulatory framework on the security of radioactive materials in Cameroon	M. FOKOU N.M.M. MAURICE S. AUGUSTIN *Cameroon National Radiation Protection Agency* Cameroon
80	A comprehensive review: interfaces between nuclear security and State systems of accounting and control	M. TEFERA KEBEDE *Ethiopian Technology Authority* Ethiopia
82	Ok. We have a received a DBT. What's next? An operator perspective	A.B. KEIZER *Urenco Nederland B.V.* Kingdom of the Netherlands
83	Why KPIs make sense and how to develop them? An operator perspective	A.B. KEIZER *Urenco Nederland B.V.* Kingdom of the Netherlands
84	Protecting nuclear power plants from social engineering	A. ARREOLA GARCÍA *National Autonomous University of Mexico — Anáhuac Mexico University* Mexico
85	Combined national retreat with remote expert assistance approaches in the development of nuclear security regulations (a successful milestone in the RURA/US-NRC partnership)	R. BANA *Rwanda Utilities Regulatory Authority* Rwanda
89	A model for predictive assessment of insider threats in nuclear facilities	M. ISMINI GALANOPOULOU A. SFETSOS I. TSOUROUNAKIS M. VARVAYIANNI *National Centre for Scientific Research Demokritos* Greece

Abstract ID	Title of paper	Contributor(s) and affiliation(s)
91	Review on detection framework for mitigating CFSI risk at Indonesian nuclear facilities	YAZIZ HASAN FADLY PUTRA JAYA JUMADIONO JUMADIONO IWAN HERU PURNOMO DWI RAHAYU HANNA YASMINE NANA SUPRIATNA ARIEF SASONGKO ADHI ALIM MARDHI TEGUH ASMORO ADE RUSTIADAM SUKONO SUKONO *National Research and Innovation Agency* Indonesia
94	Kazakhstan's nuclear security training center: International cooperation enhancing national and regional training capabilities	N. IZMAILOVA M. IDRISSOVA *Institute of Nuclear Physics* Kazakhstan
95	Design and implementation of a national framework for managing response to nuclear security events in Nigeria — Challenges and prospects	I. SAMBO *Nigerian Nuclear Regulatory Authority* Nigeria
102	Advancing isotope identification algorithms for mid-resolution detectors	GANG LI J. ALEXANDER M. ECHLIN K. STOEV M. THOMPSON *Canadian Nuclear Laboratories* Canada
103	Türkiye's experience on development of nuclear security regime	B. AKBAY T.H. MERMER S.T. ECEVİT *Nuclear Regulatory Authority* Türkiye
104	Storm series nuclear security exercise: Introduction and lessons learned	BO ZONG ZIPING LI ZHIYONG WANG FANGLEI CHEN DA LI CHANGJIE YANG SAI WANG *State Nuclear Security Technology Center* China
105	An overview of CNSC cyber security inspections at high security nuclear facilities	CUHL HWAN JUNG YUBO LEI J. SLADEK J. SIGETICH *Canadian Nuclear Safety Commission* Canada

Abstract ID	Title of paper	Contributor(s) and affiliation(s)
106	RACVIAC — US regional cooperation to strengthen nuclear security in southeastern Europe	B. ROTIM *RACVIAC Centre for Security Cooperation* Croatia J. CONNER *Los Alamos National Laboratory* United States of America P. O'BRIEN *United States Department of Energy, National Nuclear Security Administration* United States of America
111	Systematic analysis of developments of IEMI-tools with potential threats for nuclear security	M. GEHRKE M. PELZER *Installation and Reactor Safety Company (GRS)* Germany
112	Next steps for the Amendment to the Convention on the Physical Protection of Nuclear Material (A/CPPNM): How the next Conference of the Parties to the A/CPPNM can be even better than the last	S. LACKNER *Vienna Center for Disarmament and Non-proliferation* Austria
114	Break-in delay assessment: Using legacy to shape the future of high activity sources security	T. LANGUIN *Ministry of Ecological Transition and Territorial Cohesion* France O. FICHOT S. MUSSARD *Nuclear Safety and Radiation Protection Authority (ASNR) (formerly the Institute for Radiological Protection and Nuclear Safety)* France
117	Nuclear security training and qualification challenges: A case study for technical service organization in Ghana	F. OTOO *Ghana Atomic Energy Commission* Ghana S. ADU *Nuclear Regulatory Authority* Ghana P. ATTA AMOAH *Ghana Atomic Energy Commission* Ghana

Abstract ID	Title of paper	Contributor(s) and affiliation(s)
123	Japan–US cooperation on development of a high-activity solid waste measurement system during decommissioning of Tokai reprocessing plant in support of nuclear material accounting and control for nuclear security purposes	HIRONOBU NAKAMURA TAKAHIKO KITAO NORIKO MIYAJI *Japan Atomic Energy Agency* Japan J. CONNER A. LAFLEUR M. WATSON *Los Alamos National Laboratory* United States of America A. IYENGAR *United States Department of Energy, National Nuclear Security Administration* United States of America
124	Detection and interdiction of nuclear and other radioactive materials for a safer world	E. MAYAKA *Kenya Nuclear Regulatory Authority* Kenya
126	Regulation on counter-drone security system at nuclear power plants in China	LIANG ZHANG *Nuclear and Radiation Safety Center* China
128	Role of human resources capability in sustaining nuclear security detection architecture in Zimbabwe	A. MUZONGOMERWA *Radiation Protection Authority of Zimbabwe* Zimbabwe
129	International Nuclear Security Education Network (INSEN): Building capacity for youths in nuclear security	A. GOEL *Amity University* India A. KUYE *University of Port Harcourt* Nigeria M. GERLINI *University of Siena* Italy

Abstract ID	Title of paper	Contributor(s) and affiliation(s)
131	Nuclear security capacity-building: Adapting criteria for educational programs and activities	H. HALL *University of Tennessee — Knoxville* United States of America J. HARRIS *Purdue University* United States of America M. KALININA-POHL *James Martin Center for Nonproliferation Studies* United States of America C. MARIANNO *Texas A&M University* United States of America C.R.E. DE OLIVEIRA *University of New Mexico* United States of America
132	Capacity building partnerships in international nuclear security	P. ZAHNLE *Sandia National Laboratories* United States of America A. LOBATO *Army Engineering Institute* Brazil
136	Integrating Mobile-Integrated Nuclear Security Network (M-INSN) for enhanced radiation detection and response systems in Thailand through the scientific and technical support functions of the National Nuclear Security Training and Support Centre (NSSC)	ISSARIYA CHAIRAM AKKALUK CHAIWAT PRANNICHA HONGPITAKPONG THEERAPATT MANUWONG THITIDEJ TULARAK *Ministry of Higher Education, Science, Research and Innovation* Thailand
140	Canadian efforts to modernize the nuclear security regulatory framework	R. DUGUAY *Canadian Nuclear Safety Commission* Canada
142	Modelling tools for enhancing nuclear border security	G. HARRISSON J. ALEXANDER A. ERLANDSON D. GODIN O. KAMAEV A. MAUNSELL D. PÉREZ-LOUREIRO E.T. RAND D. THAKKAR M. THOMPSON J. WOO *Canadian Nuclear Laboratories* Canada

Abstract ID	Title of paper	Contributor(s) and affiliation(s)
143	A focus on non-destructive testing applied to nuclear forensics	M. TOTLAND A. AYYAGARI D. CLUFF I. DIMAYUGA T. DOMINGO A. FOURNY R. OSMOND D. THOMPSON *Canadian Nuclear Laboratories* Canada
150	Vulnerabilities determination of the physical protection elements and risk assessment using SAVI, and DAI et al model	A. AMIR *Egyptian Atomic Energy Authority* Egypt
152	Good practices in nuclear security regarding radioactive material out of regulatory control — Case studies	A. IOANNIDOU G. KITIS *Aristotle University of Thessaloniki* Greece
157	3S interface considerations for small modular reactors	P. KARHU T. HONKAMAA M. TUOMAINEN *Radiation and Nuclear Safety Authority* Finland
158	Nuclear security capacity building in Tajikistan: Challenges and ways forward	I. MIRSAIDZODA *Nuclear and Radiation Safety Agency* Tajikistan O. AZIZOV *Chemical, Biological, Radiological, and Nuclear Safety and Security Agency* Tajikistan
159	Interfaces between nuclear security and state systems of accounting and control: The US nuclear materials management and safeguards system	A. AL-DAOUK L. BEGLEY *United States Department of Energy, National Nuclear Security Administration* United States of America
162	Viet Nam's experience with INSSERV mission	NU HOAI VI NGUYEN TUAN KHAI NGUYEN THI THUY ANH BUI *Viet Nam Agency for Nuclear and Radiation Safety* Viet Nam
163	Considerations for improving public communication in nuclear security measures for public events (the case for Uganda)	L. KHALAYI *Atomic Energy Council* Uganda
165	Challenges and controversies of artificial intelligence for nuclear security	H.I. SALEH A. ARAFA *Egyptian Atomic Energy Authority* Egypt

Abstract ID	Title of paper	Contributor(s) and affiliation(s)
166	Capacity building to achieve sustainable nuclear security education and training in Myanmar	SAW THANTAR *Technological University* Myanmar
174	I&C systems less susceptible to cyberattacks	R. ARIANS L. KLEINERT O. REST A. SCHUG B. ULRICH *Installation and Reactor Safety Company (GRS)* Germany
176	Challenges, gaps, and recommendations for incursions from uncrewed aircraft systems (UAS) at critical infrastructure	O. ROSEN *Israel Atomic Energy Commission* Israel C. BURR M. POTTER *Sandia National Laboratories* United States of America
177	From marginal to mainstream: Uganda's efforts in integrating nuclear security into the mainstream national security framework	N.D. LUWALIRA *Atomic Energy Council* Uganda
178	GCNEP: India's centre of excellence	S.K. AGRAWAL *Department of Atomic Energy* India
180	Nuclear security regulations and legislation in Uganda	E. RWOTHUMIO *Atomic Energy Council* Uganda
183	Perceptions of gender balance and inclusivity by female students and trainees in the global nuclear sector	M. MALLARI F. MAHER *Organisation for Economic Co-operation and Development Nuclear Energy Agency*
188	Development of a neutron detector as an array of crystalline organic scintillators	L. LOPEZ G. LELAIZANT B. LAURENT *French Alternative Energies and Atomic Energy Commission* France X. LEDOUX *Large Heavy Ion National Accelerator* France
190	The road to the first force-on-force exercise in Hungary	A.M. VIPLAK Z. STEFÁNKA A. JEZERI *Hungarian Atomic Energy Authority* Hungary

Abstract ID	Title of paper	Contributor(s) and affiliation(s)
192	Supply chain risks in digital communication standards for nuclear safety and security	A. SCHUG L. KLEINERT O. REST *Installation and Reactor Safety Company (GRS)* Germany
195	Methods of protection against modern, high-tech computer attacks on automated systems of nuclear facilities	S.I. ZHURIN *JSC Federal Center for Science and High Technology Special Scientific and Production Enterprise (Eleron)* Russian Federation
196	A study on how nuclear material accountancy and control measures can counter an insider threat	I. TSOUROUNAKIS A. SFETSOS M. GALANOPOULOU M. VARVAYIANNI *National Centre for Scientific Research Demokritos* Greece
198	A method to distinguish background neutron sources from sources out of regulatory control using a mobile radiation search system	C.M. MARIANNO *Texas A&M University* United States of America
201	National experience on the implementation of the regulatory requirements of the security of radioactive materials in Saudi Arabia	B.A. BAHAI N. ALSULAMI S. ALSULTAN Saudi Arabia
202	Introducing an artificial intelligence as a service (AIAAS) platform for nuclear security via gamma ray monitoring systems data analysis	M. MAHMOODVAND *Atomic Energy Organization of Iran* Islamic Republic of Iran
206	Kenya's regulatory framework for security of nuclear and radioactive materials during transport	I.W. MUNDIA J.K. CHUMBA J.W. WAKUYU *Kenya Nuclear Regulatory Authority* Kenya
207	Security challenges due to the appearance of counterfeit, fraudulent and suspect items in procurement process of the nuclear supply chain	M. ĆUJIĆ M. RADENKOVIĆ L. JANKOVIĆ MANDIĆ *University of Belgrade* Serbia
211	Eliminating the risk of high activity radioactive sources by preventing adoption of Cs-137 based blood irradiation: A project to provide X ray based blood irradiation to the Joint Clinical Research Center in Kampala, Uganda	L. NAHWERA *Joint Clinical Research Centre* Uganda J. LIEBERMAN *Sandia National Laboratories* United States of America C. KITYO MUTULUUZA S. FRANCIS *Joint Clinical Research Centre* Uganda

Abstract ID	Title of paper	Contributor(s) and affiliation(s)
212	Nuclear security status for high category sources and legal framework in Ethiopia	B.T. TOLAWAK *Ethiopian Technology Authority* Ethiopia
216	Dynamic threat assessment for enhanced nuclear transport security: A collaborative AI-based approach	A. TOUARSI A. KHARCHAF *University of Ibn Tofail* Morocco
218	Incorporating diversity, equity, inclusion, and accessibility in nuclear security culture through professional regional networks	O. MARTIN *Los Alamos National Laboratory* United States of America M. KALININA-POHL *James Martin Center for Nonproliferation Studies* United States of America F. OZGE KARAKAS *Istanbul Technical University* Türkiye
219	Ways to realize gender equality in nuclear security focused on the case of the Republic of Korea	EUNBEE PARK SUNG YOON PARK *Korea Institute of Nuclear Nonproliferation and Control* Republic of Korea
224	An intentional nuclear forensics approach to support nuclear security: The use of taggants to support rapid provenance assessment	N.E. MARKS *Lawrence Livermore National Laboratory* United States of America R.M. CHAMBERLIN *Los Alamos National Laboratory* United States of America M. WELLONS *Savannah River National Laboratory* United States of America A.E. SHIELDS *Oak Ridge National Laboratory* United States of America
227	How the IPPAS mission contributed to strengthening the security system in the Czech Republic	E. KLOBOUCEK *State Office for Nuclear Safety* Czech Republic
230	An Australian perspective on the assessment of nuclear security culture for a research reactor when conducted as part of a periodic safety and security review (PSSR)	T. VAN DER VELDEN C. RISTEVSKI *Australian Nuclear Science and Technology Organisation* Australia

Abstract ID	Title of paper	Contributor(s) and affiliation(s)
231	International data sharing for modeling and simulation of physical protection system improvements	T. VAN DER VELDEN A. VIPOND *Australian Nuclear Science and Technology Organisation* Australia A. ORR B. STROMBERG *Sandia National Laboratories* United States of America
233	The development of a virtual training workshop on nuclear security measures for major public events during COVID	R.J. MAURER S. MUKHOPADHYAY *United States Department of Energy, National Nuclear Security Administration* United States of America
234	A differentiated and targeted model of security culture for radioactive sources	I. KHRIPUNOV *University of Georgia* United States of America C. SPEICHER *Ministry for the Environment, Climate and Energy Sector* Germany
239	Global Nuclear Safety and Security Institute (GNSSI) — Practice-oriented training of Russian and foreign students on nuclear security, taking into account an integrated approach to security	E. BOLOGOV N. ZHDANOVA L. NIKOLAEV *State Atomic Energy Corporation 'Rosatom'* Russian Federation
240	Methods for cybersecurity risk assessment in nuclear facilities: Adherence to regulatory requirements	N. AGBEMAVA E. MENSAH *Nuclear Regulatory Authority* Ghana
243	Enhancing supply chain cyber assurance in the nuclear sector: Findings from a comprehensive workshop series	S.N. MELEKWE P. SMITH M. ROUNCEFIELD *Lancaster University* United Kingdom
245	Practices and suggestions on nuclear security for major public events	SHUO WANG HONG LU CHEN CHEN HUIRONG WANG *State Nuclear Security Technology Center* China

Abstract ID	Title of paper	Contributor(s) and affiliation(s)
248	US Department of Energy security risk assessment overall security risk methodology	S.N. CALLAHAN M. SPARKS M. HOJNACKE G. WHITE E. ALAS *United States Department of Energy* United States of America M. PINCOCK *Pacific Northwest National Laboratory* United States of America J. SANDOVAL *Sandia National Laboratories* United States of America
250	The US Department of Energy's use of performance testing to evaluate the effectiveness of physical security systems	C.A. POCRATSKY D. GOLDEN *United States Department of Energy* United States of America J. SANDOVAL *Sandia National Laboratories* United States of America M. SPARKS K. WEBBER *United States Department of Energy* United States of America
253	40 years of monitoring the transport of nuclear materials in France	M.P. BOYER S. EVO A. MONTMEAT *Nuclear Safety and Radiation Protection Authority (ASNR)* France
254	Leveraging international partnerships to foster evolving radiological source security	J. MANION *Pacific Northwest National Laboratory* United States of America BUI THI THUY ANH *Viet Nam Agency for Radiation and Nuclear Safety* Viet Nam J. BURNS *Pacific Northwest National Laboratory* United States of America
260	Modernisation of the nuclear security framework for SMRs — Theft and sabotage	K. HEPPELL-MASYS Canada A. NILSSON Sweden

Abstract ID	Title of paper	Contributor(s) and affiliation(s)
261	IAEA regional exercises as a mechanism of increasing of national capacity of radiological incidents investigations	V.A. STEBELKOV *Laboratory for Microparticle Analysis* Russian Federation
262	Capacity building: Fostering the next generation of nuclear security experts with an eye on diversity	J. VALENCIA *Pacific Northwest National Laboratory* United States of America J. ARAPOFF *United States Department of Energy, National Nuclear Security Administration* United States of America
263	Assessing the maturity of cybersecurity and nuclear security programs at nuclear facilities	C.S. GLANTZ J.L. KNIGHT J.P. LOFTUS E.G. MCNALLY C.P. TEBBS K.D. CORNELISON A. JOSHI S.L. MORRIS S. BHADRA G.L. KIDD J. APPIAH-KUBI R.J. HARRINGTON A.P. ARENDS B. DEAN C.S. NG *Pacific Northwest National Laboratory* United States of America
264	User-friendly software for assessing the maturity of supply chain security and secure design and development programs for hardware and software at nuclear facilities	C.S. GLANTZ S.R. MIX J. APPIAH-KUBI J.P. LOFTUS E.G. MCNALLY C.P. TEBBS A. JOSHI G.L. KIDD R.J. HARRINGTON J.L. KNIGHT A.P. ARENDS B. DEAN C.S. NG K.D. CORNELISON J. MARTINOV V. REPALLE M. SARMA *Pacific Northwest National Laboratory* United States of America

Abstract ID	Title of paper	Contributor(s) and affiliation(s)
266	Challenges and opportunities in maintaining dependable nuclear communication framework that shape future collaborative public communication and stakeholder confidence: A case review of Ethiopia	A. TIRUNEH *Ethiopian Technology Authority* Ethiopia
269	Development and application of personnel reliability assessment system in nuclear security	CHEN CHEN HONG LU SHUO WANG ZIPING LI *State Nuclear Security Technology Center* China
270	How can we enhance nuclear security through an innovative methodology of observation and shock wave tracking in harsh environment?	S. TRÉLAT C. MANO *Nuclear Safety and Radiation Protection Authority (ASNR)* France M.O. STURTZER J.-F. LEGENDRE *French–German Research Institute of Saint-Louis* France
271	French centralized nuclear material accounting: A tool at the heart of national security and safeguards implementation	R.B. NGUYEN C. JACQUELIN *Nuclear Safety and Radiation Protection Authority (ASNR)* France
279	Optimization of surveillance camera site locations and viewing angles for improved nuclear security	A. SALMAN *Egyptian Atomic Energy Authority* Egypt
282	This first self-assessment of nuclear security culture in the Polish medical facility that uses radioactive sources and materials	M. WISNIEWSKA *Poznan University of Technology* Poland

Abstract ID	Title of paper	Contributor(s) and affiliation(s)
284	Jointly exercising low-visibility law enforcement nuclear security mission with practitioners from Czechia, Hungary, and the United States	N. KOCSIS *Counter Terrorism Centre* Hungary T. HOLUB *National Counterterrorism, Extremism and Cybercrime Agency* Czech Republic B. MARSI *Counter Terrorism Centre* Hungary M. PAFF *Culmen International LLC* Brazil B. SHULMAN A. PIDLUSKY *United States Department of Energy* United States of America
287	The United States Department of Energy/National Nuclear Security Administration's Office of Radiological Security program's high-risk response workshop pilot with the Republic of Moldova	A. PARNELL *Pacific Northwest National Laboratory* United States of America I. GISCA *Moldovan National Radioactive Waste Management Company* Republic of Moldova J. ANDERSON S. PUDROSCHI *Pacific Northwest National Laboratory* United States of America
289	Risk assessment in the nuclear security of natural uranium	E. NIKHANOV *Kazatomprom* Kazakhstan
297	Addressing insider threats in transporting radioactive materials	K. KALDENBACH B. LOFTIN *Oak Ridge National Laboratory* United States of America N. UYS D. SEEPAMORE *South African Health Products Regulatory Authority* South Africa

Abstract ID	Title of paper	Contributor(s) and affiliation(s)
298	Increasing maintenance management capabilities through collaborative workshops: Jordan's solution for a maintenance management system in cooperation with the United States	L.A. KHALIL A. RASHAIDEH *Energy and Minerals Regulatory Commission* Jordan J. MESSIMORE *Oak Ridge National Laboratory* United States of America G.A. ADAMS *Pacific Northwest National Laboratory* United States of America
299	Computer security regulations for nuclear facilities tailored for Indonesia needs	L.S. SETIANINGSIH *Nuclear Energy Regulatory Agency* Indonesia D. CHRISTENSEN J.L. KNIGHT G.P. LANDINE *Pacific Northwest National Laboratory* United States of America DJOKO HARI NUGROHO *Nuclear Energy Regulatory Agency* Indonesia
302	Management of an aging workforce in the nuclear security sector	FADLY PUTRA JAYA JUMAIONO YAZIZ HASAN HERI GUNAWAN *National Research and Innovation Agency* Indonesia
303	Pakistan's experience for establishing and implementing nuclear security regime: Lessons learnt for embarking countries	N. IFTAKHAR A. SHAKOOR *Pakistan Nuclear Regulatory Authority* Pakistan
308	The United Nations Office on Drugs and Crime (UNODC) repository of national legislation for implementing the criminalization provisions of the International Convention for the Suppression of Acts of Nuclear Terrorism (ICSANT), the Convention for the Physical Protection of Nuclear Material (CPPNM) and its 2005 Amendment (A/CPPNM)	M. LORENZO SOBRADO M. REGGI *United Nations Office on Drugs and Crime*
310	Support for the nuclear security component in the IRSN crisis management organization	W. BOUCLIER *Nuclear Safety and Radiation Protection Authority (ASNR)* France
313	Strengthening international cooperation through the International Convention for the Suppression of Acts of Nuclear Terrorism	F. ANDRIAN A. LAZAREV *United Nations Office on Drugs and Crime*

Abstract ID	Title of paper	Contributor(s) and affiliation(s)
315	Approach for management and sustainability of national nuclear detection architecture	A. ULLAH S. HUSSAIN *Pakistan Nuclear Regulatory Authority* Pakistan
321	United States Nuclear Regulatory Commission assessment of emerging technologies and threats in nuclear security	A. BROWN A. KIM K. LAWSON-JENKINS I. GARCIA S. PRASAD *United States Nuclear Regulatory Commission* United States of America
322	Designing a strategy (threat/target sets/denial/defense in depth)	J.K. CLARK *United States Nuclear Regulatory Commission* United States of America
323	Cybersecurity considerations of autonomy in nuclear facilities (cybersecurity licensing activities)	A. KIM K. LAWSON-JENKINS D. ESKINS I. GARCIA T. RIVERA *United States Nuclear Regulatory Commission* United States of America
324	Hostile action-based drill and exercise lessons inform emergency preparedness for security program effectiveness	C. ROSALES-COOPER *United States Nuclear Regulatory Commission* United States of America
325	Update to NRC decommissioning activities related to nuclear security	V. WILLIAMS S. COKER B. ZALESKI *United States Nuclear Regulatory Commission* United States of America
326	A novel fitness for duty approach for advanced reactors	B. ZALESKI *United States Nuclear Regulatory Commission* United States of America
330	Security for new reactors	T.R. LEACH *United States Nuclear Regulatory Commission* United States of America
331	NRC regulatory efforts for cybersecurity of advanced reactors	T. RIVERA I. GARCIA *United States Nuclear Regulatory Commission* United States of America
332	Updates to, and application of, IAEA TDL-006 and ORS best practices for computer security assessments of other radioactive material sites	J. DECASTRO M. ROWLAND *Sandia National Laboratories* United States of America T. NELSON *International Atomic Energy Agency*
335	HEU minimization in the 21st century: A shifting landscape	J. CHAMBERLIN *United States Department of Energy, National Nuclear Security Administration* United States of America

Abstract ID	Title of paper	Contributor(s) and affiliation(s)
337	Achieving gender parity in a State's nuclear security regime: A case study of the Philippines	J.L. SABLAY *Philippine Nuclear Research Institute* Philippines C. CHAUVET-MALDONADO M. A. FELDSTEIN *Oak Ridge National Laboratory* United States of America M.T. SALABIT *Philippine Nuclear Research Institute* Philippines
338	Nuclear security policy: Case study perspectives on the development and implementation in Colombia, Kenya, and Nigeria	C. CHAUVET-MALDONADO M. MAN *Pacific Northwest National Laboratory* United States of America N. BELLO *Nigerian Nuclear Regulatory Authority* Nigeria A. GALGALLO D. MBUGUA *Nuclear Power and Energy Agency* Kenya J.M. VILANUEVA *Ministry of Mines and Energy* Colombia
346	Regulatory competence management for nuclear security	A. SHAKOOR *Pakistan Nuclear Regulatory Authority* Pakistan
351	Deterrence: Evolving threats and approaches to strengthen nuclear security	R.M. RODGER H. DOWSETT *United Kingdom National Nuclear Laboratory* United Kingdom
356	How to sleep soundly whilst balancing secure by design and operational challenges	H. DOWSETT S. PROUD *United Kingdom National Nuclear Laboratory* United Kingdom
357	Enhancing computer security of nuclear power plants based on explainable AI	JIANGHI LI M.U. OZBEK *Tsinghua University* China
358	Interested parties involvement in development of nuclear security regulations in Ghana	E. AMPOMAH-AMOAKO S. ADU N. ALLOTEY *Nuclear Regulatory Authority* Ghana

Abstract ID	Title of paper	Contributor(s) and affiliation(s)
359	Nuclear security cooperation among nuclear regulators in Africa	E. AMPOMAH-AMOAKO *Nuclear Regulatory Authority* Ghana J. CHIPURU *Radiation Protection Authority of Zimbabwe* Zimbabwe LINGQUAN GUO *International Atomic Energy Agency* Y. IDRIS *Nigerian Nuclear Regulatory Authority* Nigeria INSUK JANG *International Atomic Energy Agency* A. LASFAR *Moroccan Agency for Nuclear and Radiological Safety and Security* Morocco S. SHAABAN *Egyptian Nuclear and Radiological Regulatory Authority* Egypt
360	Radioactive sealed source removal from a Guatemalan hospital	R.S. BRIDGES *Defense Threat Reduction Agency* United States of America M. WALD-HOPKINS *Los Alamos National Laboratory* United States of America F. REYES *National Institute of Oncology* Guatemala B. RABAEY *United States Department of Energy, National Nuclear Security Administration* United States of America
365	Floating nuclear power plants security: Potential threats and risks	BAMBANG TRI PURNOMO LUKMAN HAKIM *Nuclear Energy Regulatory Agency* Indonesia
366	Key factors for sustaining nuclear security detection architecture in Argentina: Roadmap for a legitimate and effective architecture	F. CIAMPAGNA MOLINA *National Ministry of Security* Argentina

Abstract ID	Title of paper	Contributor(s) and affiliation(s)
367	The graduate nuclear security program at the Kyiv Polytechnic Institute: Status update, insights and next steps	T. BIBIK *National Technical University of Ukraine* Ukraine A.D. WILLIAMS *Sandia National Laboratories* United States of America
368	Prepare to be boarded: Cooperatively developing international maritime counter nuclear smuggling capabilities	U. KHAN *United Nations Office on Drugs and Crime* P. TRASK *Sandia National Laboratories* United States of America
371	Harmonization of nuclear-security rules and regulations for maritime nuclear systems between International Maritime Organization and International Atomic Energy Agency	JORSHAN CHOI United States of America
375	A regulatory approach to sabotage of nuclear material transports for densely populated areas	B. MASSING T. PISSULLA *Federal Ministry for the Environment, Nature Conservation, Nuclear Safety and Consumer Protection* Germany
377	The legal and regulatory framework for nuclear security in India	A. KUMAR G. SHARMA A. KUMAR NAYAK *Department of Atomic Energy* India
378	Russian view on effective national physical protection regime	R. BAYCHURIN S. MAROGULOV *State Atomic Energy Corporation 'Rosatom'* Russian Federation
380	PNRA efforts for the security of nuclear and other radioactive material out of regulatory control (MORC)	S. ZEESHAN H. YOUNIS M. YAHYA RIAZ *Pakistan Nuclear Regulatory Authority* Pakistan
391	Communicating nuclear security to public through risk communication concept	RETNO AGUSTYAH *Nuclear Energy Regulatory Agency* Indonesia
392	Strengthening the regulatory framework for nuclear security in Cameroon: Difficulties encountered, results and prospects	H.J. NZOUATCHA L.L. MOHAMADOU A. SIMO *National Radiation Protection Agency* Cameroon
393	Assessment of the transport security requirements in Indonesian government regulation on radiation safety and security on transport of radioactive materials	VATIMAH ZAHRAWATI *Nuclear Energy Regulatory Agency* Indonesia

Abstract ID	Title of paper	Contributor(s) and affiliation(s)
394	Experience in the development of procedures for the assessment of the borehole disposal facility	MUHAMMAD HASSYAKIRIN HASIM MONALIJA KOSTOR MOHD ZULFADLI RAMLI NORAINI RAZALI *Department of Atomic Energy* Malaysia
395	INSEN: Shaping the future of nuclear security education	W. METWALLY *Oak Ridge National Laboratory* United States of America M. SHAFIQUL ISLAM *University of Dhaka* Bangladesh C. FORD *United States Department of Energy, National Nuclear Security Administration* United States of America
396	Shaping the future: Building needed dynamism into the nuclear security regime	K. BRILL *The Henry L. Stimson Center* United States of America J. BERNHARD *International Nuclear Security Forum* Denmark
399	Helping to shape the future: The UK's National Centre for Nuclear Security and Non-Proliferation	R. HOWSLEY J. EDWARDS M. BUDSWORTH *United Kingdom National Nuclear Laboratory* United Kingdom
403	Nuclear security activities during and after the COVID-19 pandemic: The ARN's experience	M.A. TAPIA G.M. ACOSTA P.M. ZUNINO *Nuclear Regulatory Authority* Argentina
405	Pakistan's Centre of Excellence in Nuclear Security: The vehicle for capacity building and international cooperation	M. SHAHZAD M. HUSSAIN *Pakistan Atomic Energy Commission* Pakistan
407	A retrospective assessment of the role of key universities in new nuclear power programmes: Nairobi's decade engagement in capacity building in nuclear security and forensics research	H.K. ANGEYO *University of Nairobi* Kenya
411	Mechanism of communication among response organizations and with public in case of nuclear security events	B. MUSHTAQ R. MAJEED S. JAVAID *Pakistan Atomic Energy Commission* Pakistan

Abstract ID	Title of paper	Contributor(s) and affiliation(s)
412	Integration of the response to nuclear security events into the German CBRN response capabilities at the federal level (CBRN support network)	E.A. KROEGER A. RUPP M. BARON J.-T. EISHEH E. GERICKE B. LANGE J. VOGT *Federal Office for Radiation Protection* Germany
413	German federal crime scene exercise in May 2023	E.A. KROEGER M. BARON J.-T. EISHEH E. GERICKE B. LANGE A. RUPP J. VOGT *Federal Office for Radiation Protection* Germany
414	Analysis of Chinese practice in fulfilling international nuclear security obligations	CHUNHUA QIU LIMING WANG *State Nuclear Security Technology Center* China
416	Regulatory approach for management of nuclear safety–security interface	H. YOUNIS A. MUHAMMAD *Pakistan Nuclear Regulatory Authority* Pakistan
417	Cyber-attack on nuclear regulatory system servers	RAJA ABDUL AZIZ RAJA ADNAN *Atomic Energy Licensing Board* Malaysia NORAISHAH PUNGGUT MOHD IRWAN EFFENDIN MOHD NORDIN *Department of Atomic Energy* Malaysia

Abstract ID	Title of paper	Contributor(s) and affiliation(s)
		V. DEROUET *Électricité de France (EDF)* France
		J. BOUYER P. YVON *French Alternative Energies and Atomic Energy Commission* France
423	French COE: A tool for capacity building for nuclear security	C. CORNAND *Framatome* France
		L. ISNARD *Orano* France
		G. SAQUET *National Radioactive Waste Management Agency (Andra)* France
		H. KRÖGER J. EISHEH *Federal Office for Radiation Protection* Germany
424	Theft of Ni-63 and Pu-238: Lessons learned from the investigation	C. VIDAL K. GRAHAM *Criminal Police Office Lower Saxony* Germany
		M. KNAUER *Lower Saxony Water Management, Coastal and Nature Protection Agency* Germany

Abstract ID	Title of paper	Contributor(s) and affiliation(s)
425	Strengthening nuclear security capacity in Thailand and the ASEAN region to meet the challenges of the future	ISSARIYA CHAIRAM *Ministry of Higher Education, Science, Research and Innovation* Thailand S.A. CARLSON *World Institute for Nuclear Security* Austria THITIDEJ TULARAK KAMOLPORN PAKDEE SUCHAYA KAJONCHOTIPHONG *Ministry of Higher Education, Science, Research and Innovation* Thailand R. DELGADO *World Institute for Nuclear Security* Austria
426	Cyber and physical security: Far away, so close	P.J. VENEMA *Urenco Nederland B.V.* Kingdom of the Netherlands
427	The quantification of security risk assessment	G.R. FOSTER J. EDWARDS *United Kingdom National Nuclear Laboratory* United Kingdom
429	Revolutionising threat analysis using new media technology	A. SHRIVASTAVA *World Institute for Nuclear Security* Austria
430	Experience sharing from implementing the alternative technology in radiotherapy in Tunisia: Challenges and opportunities for radiological security	L. BOUGUERRA BEN OMRANE *National Centre of Radiation Protection* Tunisia J. DAOUD I. MELLAKH *Ministry of Health* Tunisia L. SPANGLER *Sandia National Laboratories* United States of America

Abstract ID	Title of paper	Contributor(s) and affiliation(s)
431	Toward gender equality, diversity and inclusion as a core nuclear security principle	R. EVANS *World Institute for Nuclear Security* Austria P. NAIDOO AMEGLIO *Australian Nuclear Science and Technology Organisation* Australia K. PAUL *Bruce Power* Canada L. AMUR V. ESPINOSA MORENO *World Institute for Nuclear Security* Austria
434	Am I deploying effective detection systems?	P.M. JOHNS *Pacific Northwest National Laboratory* United States of America M.J. DEMBOSKI *United States Department of Energy, National Nuclear Security Administration* United States of America
436	Advances in diversity and inclusion in radioactive material security programs	M.E. EKMAN *Sandia National Laboratories* United States of America
437	The algorithm that will eliminate millions of nuisance alarms	P. JOHNS P. DIMMERLING A. MOORE B. JENNINGS C. MATTOON S. LABOV N. ROWE M. KUHN M. DELARUBIA J. FRIEDMAN E. KUNGLAMAE *United States Department of Energy, National Nuclear Security Administration* United States of America

Abstract ID	Title of paper	Contributor(s) and affiliation(s)
438	Assessment of nuclear security culture in the transportation of radioactive material at Malaysian nuclear agency	MOHD FAZLIE ABDUL RASHID *Malaysian Nuclear Agency* Malaysia TONI WAHYU PAMUNGKAS *Nuclear Energy Regulatory Agency* Indonesia HUSAINI SALLEH *Malaysian Nuclear Agency* Malaysia
439	National framework for managing the response to nuclear security events: Cuban experience	Y.L. FORTEZA J.L.P. GILISMAN *Office for Environmental Regulation and Safety* Cuba
441	Development of a new international IEC standard: Security of medical equipment using high-activity radioactive sources	A. NILSSON Sweden M. HARTKOPF *Varian Medical Systems Inc.* United States of America G. IBBOTT *University of Texas MD Anderson Cancer Center* United States of America P. KJÄLL *Elekta Instrument AB* Sweden M. KUCA *Sandia National Laboratories* United States of America P. KUMAR *All India Institute of Medical Sciences* India R. KUMAR *Bhabha Atomic Research Centre* India
447	Proliferation resistance integration for next generation research reactors	A.S. MEEHAN K.J. SALLEE N.C. IYER B.D. WAUD J. DIX C.C. LANDERS *United States Department of Energy, National Nuclear Security Administration* United States of America

Abstract ID	Title of paper	Contributor(s) and affiliation(s)
450	Identification of competencies for law enforcement roles in nuclear security	S.M. PEPER J. BUCHANAN *International Criminal Police Organization (INTERPOL)* France
461	Multifaceted challenges and opportunities associated with small modular reactors in Africa: A nonproliferation perspective	C. PRAH *Ontario Tech University* Canada S. ADU *Texas A&M University* United States of America
462	RadSecLexis — An opensource online database and tool for radiological source security frameworks	C.A. MCALLISTER A.W. TRENTHAM *The Henry L. Stimson Center* United States of America
465	Assessing progress towards compliance on the implementation of the A/CPPNM in the Philippines' nuclear security regime for nuclear material and nuclear facilities using the Oak Ridge National Laboratory (ORNL) self-assessment tool	M.T.A. SALABIT J.R.A. FERNANDEZ A.M. BORRAS *Philippine Nuclear Research Institute* Philippines
469	Emerging technologies and methods in wide-area search for nuclear materials	L.E. SINCLAIR D.A. MCCORMACK *Natural Resources Canada* Canada
477	A nuclear security regulatory model for small modular reactors	M.A. MAN P.E. EFTEKHARI F. PUTZU *Pacific Northwest National Laboratory* United States of America J. RIVERS *Sandia National Laboratory* United States of America
478	A fictional case study with a corresponding legal and regulatory framework: Challenges and good practices in the implementation CPPNM amendment article 5	M. MAN A. COOPER P.E. EFTEKHARI *Pacific Northwest National Laboratory* United States of America
479	Considerations for implementing nuclear material accounting and control used to support nuclear security into regulatory requirements: A Ghanaian experience	A. MENSAH *Nuclear Regulatory Authority* Ghana M. DOTSE *Ghana Atomic Energy Commission* Ghana M.A. MAN *Pacific Northwest National Laboratory* United States of America

Abstract ID	Title of paper	Contributor(s) and affiliation(s)
481	Radioactive source safety and security at gamma irradiation facility in the Republic of Azerbaijan	Z. KHALILOV *Ministry of Digital Development and Transport* Azerbaijan M. GAHRAMANOVA *Ministry of Emergency Situations* Azerbaijan
486	Digital radiography vehicle inspection system for access control system for physical protection of nuclear facilities	PENG CONG LITAO LI ZHENTAO WANG YANMIN ZHANG YIBIN HUANG WEIDONG QIU XIAOJING. GUO LIU CHEN GUILAI XING JIAN ZHENG LIQIANG WANG YUEWEN SUN *Tsinghua University* China
492	The Nigeria police force: Model for nuclear security detection architecture	R. OSARO UGHEGHE *Nigeria Police Force* Nigeria P. WEST *Office of the National Security Adviser* Nigeria
495	Challenge in the control of transboundary movements of portable and mobile sources	P. BOMPERE LEMO *National Committee of Protection against Ionizing Radiation* Democratic Republic of the Congo
496	Digital twin-integrated security system for nuclear reactor facility with predictive learning capabilities	R. ALFARIJI G. IVANKA H. T. L. NUGRAHA S.P. NURHIDAYAH S.H. PUTERO *University of Gadjah Mada* Indonesia
497	Theory and practice of effectiveness evaluation of physical protection system for nuclear facilities in China	MINGWEI LIU FANGLEI CHEN CHANGJIE YANG XUEMEI GAO HAN YELIANG ZIPING LI *State Nuclear Security Technology Center* China

Abstract ID	Title of paper	Contributor(s) and affiliation(s)
498	Adopting novel educational offerings (NEOS) as a means of ensuring sustainability of nuclear security educational programmes in Malaysia	IRMAN ABDUL RAHMAN RAJA ABDUL AZIZ RAJA ADNAN *National University of Malaysia* Malaysia N.F. BAKRI *Nuclear and Radiological Regulatory Commission* Saudi Arabia
499	Evaluating physical protection system for radioactive source at Universitas Gadjah Mada's Nuclear Engineering Department using stochastic approach	S.G. ISMAYA M.C.C. HUTAMA D.C. VALENTINO R.A. YASIN *University of Gadjah Mada* Indonesia
506	Importance and development of a centralized alarm assessment capacity with a central alarm station (CAS) integrated personal radiation detector network	U. LIRANAARACHCHI *Wayamba University of Sri Lanka* Sri Lanka N. RATHNAWEERA N. RANASINGHE *Sri Lanka Atomic Energy Board* Sri Lanka
514	The young generation as a key ingredient to the future of nuclear security in Uganda	R.D. MUGISA *Atomic Energy Council* Uganda
515	Sustaining human resources in the area of prevention, detection and response to nuclear and other radioactive material out of regulatory control (MORC)	R. EVANS *World Institute for Nuclear Security* Austria H. HORNE *Norwegian Radiation and Nuclear Safety Authority* Norway V. FEDCHENKO *Stockholm International Peace Research Institute* Sweden
517	Nuclear facilities protection against UAVs	J. VÁCLAV *Nuclear Regulatory Authority of the Slovak Republic* Slovakia

Abstract ID	Title of paper	Contributor(s) and affiliation(s)
519	Nuclear security and safety: A systematic review of climate-resilient measures	A.R. CARVALHO M.V. ANDRADE E.A. RODRIGUES *Nuclear and Energy Research Institute* Brazil B.R. CARVALHO *University of Brasilia* Brazil D.A. ANDRADE J.O.W.V. BUSTILLOS *Nuclear and Energy Research Institute* Brazil F. DE LOS A.R. DEL TORO *Harbin Engineering University* China
522	Argentine supply chain verifications strengthening security	L.I. VALENTINO *Nuclear Regulatory Authority* Argentina
525	Redefining the cooperative threat reduction model for multilateral nuclear security	K. TUCKER *Nuclear Threat Initiative* United States of America
526	Construction of nuclear forensic database in China	ZIPING LI JILONG ZHANG ZHIBO ZHOU LIFANG YANG *State Nuclear Security Technology Center* China TONGZING WANG *China Institute of Atomic Energy* China
536	Enhancing youth engagement in nuclear security	EUNBEE PARK SUNG YOON PARK *Korea Institute of Nuclear Nonproliferation and Control* Republic of Korea
541	Lessons learned from KINAC cyber security test-bed	IN-HYO LEE KI-HAENG NAM HYUN-JOO LEE KOOKHEI KOWN *Korea Institute of Nuclear Nonproliferation and Control* Republic of Korea

Abstract ID	Title of paper	Contributor(s) and affiliation(s)
543	Upskilling nuclear forensics at regional and international levels	J. BAGI M. WALLENIUS J. GALY Z. VARGA S. VIGIER K. MAYER *Joint Research Centre* European Commission
545	Challenges and response: China's approach to emerging threats on nuclear security	HUANG PING *State Nuclear Security Technology Center* China
547	Optimizing computer security for other radioactive material facilities	G.K. WHITE *Lawrence Livermore National Laboratory* United States of America M.T. ROWLAND J. DECASTRO *Sandia National Laboratory* United States of America T.D. NELSON *International Atomic Energy Agency*
549	Relevancy of cyber threats on nuclear transports: French studies	O. D'HÉNIN C. DEBERGE O. FICHOT *Nuclear Safety and Radiation Protection Authority (ASNR)* France C. TERTRAIS *Ministry of Ecological Transition and Territorial Cohesion* France
556	How to run remote mentoring for nuclear security practitioners	Z.S. HOMAN *King's College London* United Kingdom T. DE SCHRYVER *Dutch Defence Academy* Kingdom of the Netherlands
557	Imbibe the concept of nuclear safety and security with the aid of capacity development program: A case study of Amity Institute of Nuclear Science and Technology (AINST), Amity University Uttar Pradesh, Noida, India	A. YADAV A. DATTA U. GUPTA S. RAY A. GOEL *Amity University* India
559	Proposing a regional nuclear security mechanism	S. NOOR *Harvard University* United States of America

Abstract ID	Title of paper	Contributor(s) and affiliation(s)
560	Leveraging the interface between emergency preparedness and response and nuclear security to enhance the nuclear security regime in Uganda	J. BIRUNGI *Atomic Energy Council* Uganda
561	Strengthening nuclear security with gender-inclusive strategies in newcomer countries	C. NWAKANMA *Nigeria Atomic Energy Commission* Nigeria
566	A system for rapid detection and tracking of radioactive sources at security gates	KE LI MANCHUN LIANG SHUIJUN HE HONGMIN SHEN JIAN HUANG RUDONG WANG *Tsinghua University* China CHANGJIE YANG YELIANG HAN *State Nuclear Security Technology Center* China
567	How to take into account temporary NM detention between transports and facilities	A. BARRAULT J. LAUNAY *Department for Nuclear Security* France
571	Change management: How to deal with site evolutions? From the expectations of the competent authority (in relation to the regulations) to the implementation by operators, Example of collaborative work with Orano	A. BARRAULT A. MALABIRADE *Department for Nuclear Security* France
573	Challenges and benefits of possible use of remotely operated technologies in nuclear power plants	F. ALHAMMADI F. AL MUHAIRI *Federal Authority for Nuclear Regulation* United Arab Emirates
576	Assessing impact and sustainability of Nuclear Security Education Program at Universitas Gadjah Mada, Indonesia	S. SIHANA A. AGUNG F. FERDIANJAH G.O.F. PARIKESIT S.H. PUTERO E. WIJAYANTI *University of Gadjah Mada* Indonesia

Abstract ID	Title of paper	Contributor(s) and affiliation(s)
577	The German DBT — Challenges and progress	B. BRÜCK *Federal Ministry for the Environment, Nature Conservation, Nuclear Safety and Consumer Protection* Germany M. PELZER C. QUESTER *Installation and Reactor Safety Company (GRS)* Germany
580	Mitigating insider threats posed by domestic violent extremists	M. BUNN *Harvard University* United States of America N. ROTH *Nuclear Threat Initiative* United States of America
583	Methods to combat dynamic threat vectors	A. CHOREN *Rosca Solutions, World Nuclear Association* United States of America
585	The future is today: Virtual and augmented reality in security response training	A. ORDERS *North Carolina State University* United States of America D. HIGGINS *Pacific Northwest National Laboratory* United States of America C. LOPEZ *Sandia National Laboratories* United States of America T. WILLIAMS *Y-12 National Security Complex* United States of America
587	The NPT and nuclear security: Tracing and strengthening an important relationship	L. VAN DASSEN *World Institute for Nuclear Security* Austria M. BUDJERYN *Harvard University* United States of America W. ISAACS *World Institute for Nuclear Security* Austria
589	Regulatory aspects of decommissioning: Case study of Fessenheim NPP	A. BARRAULT *Department for Nuclear Security* France

Abstract ID	Title of paper	Contributor(s) and affiliation(s)
594	Lessons learnt from the formation of the European Nuclear Security Regulators Association (ENSRA)	R. GAUCHER *European Nuclear Security Regulators Association* France B. STAUFFER *European Nuclear Security Regulators Association* Switzerland J. CESAREK *European Nuclear Security Regulators Association* Slovenia
596	Security during the transport of nuclear and other radioactive material	Y. LIU A. BENNETT IRION *Argonne National Laboratory* United States of America J. SHENK *United States Department of Energy* United States of America
598	Development and evaluation of security culture in a research and development organization	E. GOSSET B. BOREL *French Alternative Energies and Atomic Energy Commission* France
601	Preliminary reflections from the Swedish radiation safety authority on recent AdSec/INSAG recommendations regarding safety/security interfaces	P. LINDAHL A. OBENIUS MOWITZ *Swedish Radiation Safety Authority* Sweden
604	Preparing for the second review conference on A/CPPNM — How to evaluate the adequacy of the convention in light of the then prevailing situation	P. LINDAHL *Swedish Radiation Safety Authority* Sweden
607	Development of an empirical model for evaluating the effectiveness of measures to deter potential antagonists from engaging in a malicious act	P. LINDAHL *Swedish Radiation Safety Authority* Sweden
608	Addressing the challenges of insider threats with AI	E. EIFERT D. DUDENHOEFFER *Austrian Institute of Technology* Austria
609	A simulation framework for computer security assurance activities	R. PAULINO MARQUES J.R. CASTILHO PIQUEIRA *University of São Paulo* Brazil R. BUSQUIM E SILVA *International Atomic Energy Agency*

Abstract ID	Title of paper	Contributor(s) and affiliation(s)
610	Developing cyber security operations centers (SOCs) for nuclear plant operations	E. EIFERT D. DUDENHOEFFER *Austrian Institute of Technology* Austria
611	Building capability for computer security assurance activities through international cooperation	L. MOUTENOT *Électricité de France (EDF)* France R. PAULINO MARQUES *University of São Paulo* Brazil P. SMITH *Lancaster University* United Kingdom R. BUSQUIM E SILVA *International Atomic Energy Agency*
612	Simulator of small modular reactor for cyber security assessment	R. BUSQUIM E SILVA *International Atomic Energy Agency* M. ROWLAND *Sandia National Laboratories* United States of America R. PAULINO MARQUES *University of São Paulo* Brazil
613	Severity of vulnerabilities resulting from PLC misconfigurations and developments	YOUNGSUN KIM SEUNGJUN LEE RYUNGHYE KIM *OPCIA Corp.* Republic of Korea
619	2023 nuclear security index: Falling short in a dangerous world	N. ROTH J. BROSNAN R. MATZKIN-BRIDGER *Nuclear Threat Initiative* United States of America
620	Fostering nuclear security leadership and innovation: Lessons learned from 10 years of global dialogue on nuclear security priorities	J. BUFFORD N. ROTH *Nuclear Threat Initiative* United States of America
624	Humour and nuclear security communications	S. SHAHEEN *King's College London* United Kingdom

Abstract ID	Title of paper	Contributor(s) and affiliation(s)
629	United States–Japan joint study on material attractiveness: Evaluating the malicious act risks of nuclear material theft	YOSHIKI KIMURA *Japan Atomic Energy Agency* Japan R.M. CHAMBERLIN *Los Alamos National Laboratory* United States of America J.J. BLAND *Sandia National Laboratories* United States of America
634	How far should we go: Brazilian experience on developing a national regulatory framework for computer security of nuclear facilities	R.L.A. TAVARES *Institutional Security Office* Brazil R.O. FARIA *National Nuclear Energy Commission* Brazil
635	Contributing to the advancements in civil nuclear technologies for sustainable and long term energy demands in Latin America, through capacity building and workforce development	G.M. ACOSTA *Nuclear Regulatory Authority* Argentina R.D. FLETCHER *World Institute for Nuclear Security* Austria P.C. ALIENDRE CABANAS *Radiological and Nuclear Regulatory Authority* Paraguay C. ARAUJO *Maxim Group* Brazil C.A. PRIETO VALDERRAMA *Pontifical Javieriana University* Colombia W.K. ISAACS V. ESPINOSA MORENO *World Institute for Nuclear Security* Austria
636	Virtual training strategies and lessons learned for nuclear security	G. D'ARRIGO M. STEWART-SMITH *United States Department of Energy, National Nuclear Security Administration* United States of America
638	Implementation of the nuclear security systems and security measures during the COVID-19 era	K. BRIAN *Atomic Energy Council* Uganda

Abstract ID	Title of paper	Contributor(s) and affiliation(s)
640	Computer security assessments of industrial irradiators: Lessons learned from the field	M. ROWLAND *Sandia National Laboratories* United States of America L. PEDRAZZINI *Gammatom S.R.L.* Italy M. COMBEN *International Irradiation Association* United Kingdom J. DECASTRO M. KUCA *Sandia National Laboratories* United States of America
649	Combating information disorder to address a nuclear incident	H. HARDIN *United States Department of Energy, National Nuclear Security Administration* United States of America
650	How will artificial intelligence and small modular reactors shape the future of emergency preparedness and response	R. ROCKABRAND *United States Department of Energy, National Nuclear Security Administration* United States of America
652	National legal framework of Montenegro for nuclear security, including legally binding and non-binding international instruments	T. DJUROVIC M. SRECKOVIC *Ministry of Ecology, Sustainable Development and Northern Region Development* Montenegro
653	Design and proposal of a real-time traceability and authentication system, based on blockchain and artificial intelligence techniques, for the safe transport of radioactive sources and nuclear material in Colombia	G.A. PUERTA *Ecopetrol* Colombia J.P. PARRA J.M. MARIN *Ministry of Mines and Energy* Colombia
654	Maintaining and enhancing nuclear forensics capacities	H. SCHOECH F. POINTURIER *French Alternative Energies and Atomic Energy Commission* France
659	The coordination of response to nuclear security events: Malaysia experience	DEWISINITA MOKHTAR MONALIJA KOSTOR *Atomic Energy Licensing Board* Malaysia

Abstract ID	Title of paper	Contributor(s) and affiliation(s)
661	Capacity building in Latin America and the Caribbean through the use of regional subject matter experts in transport security	P.B. LADD K.K. ANDERSON M.C. LIN *Y-12 National Security Complex* United States of America J.J. KARCZ *Oak Ridge National Laboratory* United States of America A.Z. ABADIA J.G. PHILLIPS *Secured Transportation Services, LLC* United States of America E.M. THOMPSON *United States Department of Energy, National Nuclear Security Administration* United States of America
663	Retaining a diverse nuclear workforce for the future	J. BROSNAN E. STEVENS *Nuclear Threat Initiative* United States of America
665	The future of nuclear security in Jamaica's regulatory infrastructure: Opportunities and challenges	T. WARNER *Hazardous Substances Regulatory Authority* Jamaica Z. ELLIOTT *Columbia International University, Nova Southeastern University* Jamaica
668	Lessons learned from a decade of nuclear forensics capability development in Canada	C. COCHRANE *Canadian Nuclear Safety Commission* Canada

Abstract ID	Title of paper	Contributor(s) and affiliation(s)
684	Implementation of nuclear security measures for the 50th World Petanque Championship 2023 in Benin	M.B. ZINSOU *National Authority for Radiological Safety and Radiation Protection* Benin K.S. AGOSSA T. ADJEHOUNOU K.F. LOKOSSA *Ministry of the Interior and Public Security* Benin L.R.G. AHO GLELE *Ministry of Health* Benin K.M. AMOUSSOU-GUENOU *National Authority for Radiological Safety and Radiation Protection* Benin
690	Industry's role in enhancing global nuclear security — A new model	A. BARROW *Nuclear Transport Solutions* United Kingdom
691	A regional approach to nuclear security dialogue: The Latin American case	F. DELUCHI J. GADANO *Argentina Global Foundation* Argentina N. ROTH *Nuclear Threat Initiative* United States of America
693	Compliance assurance for safe and secure transport of radioactive material vehicles licensing and inspection	S. AL BLOOSHI *Federal Authority for Nuclear Regulation* United Arab Emirates
694	Nuclear security measures for major public events in UAE	K.B. AL AMERI *Federal Authority for Nuclear Regulation* United Arab Emirates
697	Nuclear transport security in complex environments: The potential for uncrewed systems to support the security of nuclear transport operations	S. REARDON *Nuclear Transport Solutions* United Kingdom
701	The role of civil society in shaping a more secure future for all	I. KIRSTEN A.K. STOTT *Vienna Center for Disarmament and Non-Proliferation* Austria

Abstract ID	Title of paper	Contributor(s) and affiliation(s)
702	Nuclear training during a conflict: Lessons learned	B.R. BUDDEMEIER M.B. DILLON *Lawrence Livermore National Laboratory* United States of America N.F. RUSSO *Pacific Northwest National Laboratory* United States of America H. HARDIN *United States Department of Energy, National Nuclear Security Administration* United States of America
708	French regulatory guidance for SMRs	T. LANGUIN M. KOPPE *Ministry of Ecological Transition and Territorial Cohesion* France
710	Sensor evaluation and test bed facility (SETBF)	V.K. SINHA A. KUMAR V. KABRA NISHU D. PRAKASH *Global Centre for Nuclear Energy Partnership* India
718	A cross-disciplinary approach to capacity building and nuclear security	Z.D. MYERS *Defence Research and Development Canada* Canada

ANNEX III.

LIST OF FLASH PRESENTATIONS

The flash presentations[10] from the International Conference on Nuclear Security: Shaping the Future (ICONS 2024) are available on the IAEA website.

Abstract ID	Title of presentation	Contributor(s) and affiliation(s)
34	Using deep learning and Sentinel-2 satellite images for the detection of nuclear power plants	D. GARCIA DEL REAL L. GARCIA DEL REAL *Freiberg University of Mining and Technology* Germany J. GARCIA DEL REAL *Catholic University of Murcia* Spain
40	Security measures of nuclear and radioactive material handling at the seaport of Tripoli	A. SALAH ALDEEN *University of Tripoli* Libya
41	Advancing nuclear security research: Methods and tools	M.S. ISLAM *University of Dhaka* Bangladesh
53	Security and safety aspects implemented to the temporary deposit of radioactive waste (TDRW) in Mali by AMARAP	N. KONE A.A.M. DICKO A. COULIBALY S. DIARRA A.S. BERTHE *Malian Radiation Protection Agency* Mali
60	Active interrogation methods for detection and characterization of nuclear threats in enclosed structures: Challenges and mitigation schemes	G. BENTOUMI F. ALI G. LI *Canadian Nuclear Laboratories* Canada

---

[10] The flash presentations are available for consultation by registered participants of the conference either on the conference application or on the Nuclear Security Information Portal (NUSEC).

Abstract ID	Title of presentation	Contributor(s) and affiliation(s)
81	Development and evaluation of risk-informed graded approaches to ensure safe and secure trade: experiences of Chittagong seaport, Bangladesh	A.S.I. BHUIAN M. DEBNATH *Bangladesh Atomic Energy Commission* Bangladesh M.B. SAMPA *University of Science and Technology Chittagong* Bangladesh M.P. SUJAN *Megaport Initiative* Bangladesh S. HOSSAIN *Bangladesh Atomic Energy Commission* Bangladesh
99	Evaluation of the probability and impact of a possible malicious attack on the company coral FLNG platform using caesium-137 source: A risk classification approach	E. GOESSA *National Atomic Energy Agency* Mozambique
110	The nuclear security programmes: Make them more robust and inclusive by establishing a chain of capacity-building mechanisms	B. KEJELA *Ethiopian Radiation Protection Authority* Ethiopia
115	Nurturing a nuclear knowledge management culture for regulators in developing African countries to enhance nuclear security regimes	S. HUNI P. RUHUKWA *Radiation Protection Authority of Zimbabwe* Zimbabwe
116	Role of education for capacity building and stakeholder cooperation in nuclear security	Ş. UDUM *Hacettepe University* Türkiye
122	Particle production and characterization for nuclear forensics following a radiological dispersal event	A. CHAUDHURI A. AYYAGARI G. COTA-SANCHEZ I. DIMAYUGA M. MARTINEZ M. TOTLAND *Canadian Nuclear Laboratories* Canada
127	Malicious node identification in WSNS for radioactive source localization	I. MAHMOUD A. ABD EL-HAMID *Egyptian Atomic Energy Authority* Egypt
161	On-site inspections of security measures for high activity radioactive materials in Japan: Lessons learned from radioactive material security inspectors	NANA ENDO KATSUNOBU AOYAMA YUSUKE ORIBE MOTOHIRO YOSHIKAWA *Nuclear Regulation Authority* Japan

Abstract ID	Title of presentation	Contributor(s) and affiliation(s)
184	Analysis of cyber threats and methods of mitigating them at radiation hazardous facilities	Y. KHAJYNAU A. SAVIN A. DRAZHZHYNAU A. KARNILOVICH D. IVANOV M. MIKLAS *Applied Systems, Inc.* Belarus
186	Challenges faced in preparation for adherence to the Amendment to the Convention on the Physical Protection of Nuclear Material	C. GAMULANI *Atomic Energy Regulatory Authority* Malawi
204	Implementation of nuclear forensics in Thailand for supporting investigation of MORC	HARINATE MUNGPAYABAN MELASINEE LAOSUWAN AREERAK RUEANNGOEN KALAYA CHANGKRUENG PAPHON PHAUKKACHANE LADAPA SRIJITTAWA SOOWALUCK THONG-IN *Ministry of Higher Education, Science, Research and Innovation* Thailand
225	Russian approach of nuclear security and nuclear safety interface	S. MAROGULOV *State Atomic Energy Corporation 'Rosatom'* Russian Federation
228	Mitigating evolving threats and challenges to physical protection systems	M. MIRONOV *JSC SPC Dedal* Russian Federation
229	Status and considerations of adoption of X ray technologies in replacement of high risk radioactive sources	C.A. POTTER H.R. GAGARIN L.J. GILBERT D.B. HARMON R.E. MORNINGSTAR J.R. MOUSSA I.C. PADILLA S.V. RANE A. WILCOX *Sandia National Laboratories* United States of America
232	Morphological analysis for nuclear forensics signatures using SEM/EDX	AREERAK RUEANNGOEN HARINATE MUNGPAYABAN KALAYA CHANGKRUENG PAPHON PHAUKKACHANE LADAPA SRIJITTAWA SAOWALUCK THONG-IN MELASINEE LAOSUWAN *Ministry of Higher Education, Science, Research and Innovation* Thailand

Abstract ID	Title of presentation	Contributor(s) and affiliation(s)
242	Human factors in nuclear security: Lessons from the Ghana Atomic Energy Commission	R. ALGAGA M. DOTSE *Ghana Atomic Energy Commission* Ghana
265	Sustaining the nuclear security regime through the establishment and operation of national nuclear security support centres: Development and implementation of IAEA-TDL-010 methodology, challenges and lessons learnt	I. HAQ *Pakistan Institute of Engineering and Applied Sciences* Pakistan ISSARIYA CHAIRAM *Ministry of Higher Education, Science, Research and Innovation* Thailand M. SALABIT *Philippine Nuclear Research Institute* Philippines NAOKO INOUE *Japan Atomic Energy Agency* Japan M. MOHAMED GAD *Egyptian Nuclear and Radiological Regulatory Authority* Egypt Q. ROSE *International Atomic Energy Agency*
276	NNFLs as national security tools	A. STRATZ *Nuclear Security Strategies* United States of America L. DALLAS *Oak Ridge National Laboratory* United States of America
290	Enhancing physical protection measures for floating nuclear power plants	SUDI ARIYANTO AMIL MARDHA YAZIS HASAN DWI RAHAYU FADLY. PUTRA JAYA NANA SUPRIATNA *National Research and Innovation Energy Agency* Indonesia

Abstract ID	Title of presentation	Contributor(s) and affiliation(s)
292	Development of a portable field deployable physical security system and alarm monitoring station 'in a box' for capacity building program enhancement	L.R. SPANGLER M. ERDMAN M. RIMBERT C. GRAVETT *Sandia National Laboratories* United States of America T. WRIGHT *Lawrence Livermore National Laboratory* United States of America T. MACK *Sandia National Laboratories* United States of America
300	Safety and security transport of radioactive materials in Latin American region: Strategic aspects and current challenges	R. HERNANDEZ ALVAREZ *Office for Environmental Regulation and Safety* Cuba
306	Practical aspects of nuclear security for nuclear material in international transport: Experience and possible ways of improving	K. BELOUSOV *Rusatom Energy International* Russian Federation
309	Outcomes from the 7th collaborative materials exercise of the nuclear forensics international technical working group	J.M. SCHWANTES *Pennsylvania State University* United States of America O. MARSDEN *Atomic Weapons Establishment* United Kingdom
311	Ground truth measurements and forensic analysis after an explosive dispersal of radioactive materials	R.J. MAURER S. MUKHOPADHYAY *United States Department of Energy, National Nuclear Security Administration* United States of America
312	Radiation detection instruments needed for nuclear security and some NORM issues	N. HELAL E. SOBAIH *Egyptian Atomic Energy Authority* Egypt

Abstract ID	Title of presentation	Contributor(s) and affiliation(s)
		Q. SHOLLENBERGER *Lawrence Livermore National Laboratory* United States of America
		YOSHIKI KIMURA *Japan Atomic Energy Agency* Japan
		J. INGLIS *Los Alamos National Laboratory* United States of America
		R. LINDVALL *Lawrence Livermore National Laboratory* United States of America
		Y. MATSUI *Japan Atomic Energy Agency* Japan
314	US-JAEA nuclear forensics analysis to identify the origins and process history of UOCs	J. DENTON *Los Alamos National Laboratory* United States of America
		H. NISHIWAKI R. YAMANAKA K. SAWAYAMA M. NOMURA Y. UMINO *Japan Atomic Energy Agency* Japan
		N. MARKS R. KIPS *Lawrence Livermore National Laboratory* United States of America
		T. YAMAGUCHI *Japan Atomic Energy Agency* Japan
		R. STEINER *Los Alamos National Laboratory* United States of America
327	NRC overview of security training programs at nuclear power reactor sites	T.R. LEACH *United States Nuclear Regulatory Commission* United States of America
344	Policy proposal in support of nuclear security detection architecture program in Indonesia	YUDI PRAMONO EH RIYADI PETIT WIRINGGALIH *Nuclear Energy Regulatory Agency* Indonesia

Abstract ID	Title of presentation	Contributor(s) and affiliation(s)
374	Strengthening nuclear security culture at medical facilities in Malaysia: 'The second wave'	FARHANA AZIERA BADRUL HISHAM *Ministry of Health* Malaysia
400	Increasing diversity in the field of radiological security	S. PEPPER A. GONZALEZ-TORRES S. ZIA W. STERN *Brookhaven National Laboratory* United States of America K. HIRSCH *United States Department of Energy, National Nuclear Security Administration* United States of America
406	Identifying threat indicators and countermeasures	M.GUISO M. OBRIEN *Lawrence Livermore National Laboratory* United States of America
440	Comparing mobile detection systems and modular detection and identification systems: Optimizing capability for partner operations	M. DEMBOSKI R. MILLER B. SHULMAN M. SICK *Nuclear Smuggling Detection and Deterrence* United States of America C. NELSON G. PATTON *Pacific Northwest National Laboratory* United States of America J. PADILLA *Sandia National Laboratory* United States of America
446	Challenges and innovation in mobile tracking of radioactive sources in Brazil: A pioneering experience in security implementation	C. ARAUJO *Maxim Group* Brazil F.C.A. DA SILVA *Institute of Radiation Protection and Dosimetry* Brazil J. MENESES *Golden Security Services* Colombia E.S. ROJAS *Pacific Northwest Nuclear Laboratory* United States of America

Abstract ID	Title of presentation	Contributor(s) and affiliation(s)
453	Insider threat mitigation: Practical applications	Y.V. IBRAHIM A. ASUKU *Ahmadu Bello University* Nigeria S.A. DAHUNSI *Oak Ridge National Laboratory* United States of America S. JOHN S.A. JONAH *Ahmadu Bello University* Nigeria J LEWIS P.D. LYNCH *Lawrence Livermore National Laboratory* United States of America
474	Nuclear forensics implementation in the North Africa region	R. KIPS *Lawrence Livermore National Laboratory* United States of America M. IYER A. STRATZ E. WOLF *United States Department of Energy, National Nuclear Security Administration* United States of America
484	Next generation real-time 3D radiation mapping technologies for nuclear security	A. HAEFNER E. SUZUKI R. PAVLOVSKY *Gamma Reality Inc.* United States of America
513	Measuring insider risk and assessing the effectiveness of personnel security defences	S. SNOOK *Coventry University London* United Kingdom

Abstract ID	Title of presentation	Contributor(s) and affiliation(s)
		B. CARVALHO *University of Brasilia* Brazil
520	Security culture and cybersecurity in nuclear and radiological facilities: A systematic literature review	F.A.R. DEL TORO *Harbin Engineering University* China A. CARVALHO J.O.W.V. BUSTILLOS *Nuclear and Energy Research Institute* Brazil J.C. GARCIA *São Paulo State Technical School* Brazil D.A. ANDRADE E. RODRIGUES *Nuclear and Energy Research Institute* Brazil
527	Digital twins in nuclear power: Cybersecurity implications	A.S. HAHN C. LAMB J. DECASTRO M. TANAKA *Sandia National Laboratories* United States of America
535	Intelligent search and location system for nuclear and radioactive material out of regulatory control	YONG LI GUOBAO WANG YULAI ZHENG YUNTAO LIU ZIQIANG ZENG *China Institute of Atomic Energy* China
542	Research and development of a nuclear radiation tracking and positioning imaging system for nuclear security	YANCHENG YU XIN LIU ZHIHONG QIAO FENG XIE NIANMING JIANG *Beijing Novel Medical Equipment Ltd.* China QINGYANG WEI *Beijing University of Science and Technology* China TIANPENG XU YELIANG HAN *State Nuclear Security Technology Center* China

Abstract ID	Title of presentation	Contributor(s) and affiliation(s)
552	Fast and reliable nuclear threat identification in interdiction and global nuclear security, innovation using qualified technology applied to nuclear safeguards to analyze special nuclear material	M. MORICHI *CAEN* Italy
553	30 years of nuclear forensic science at the EC-joint research centre — Lessons learned and path forward	K. MAYER M. WALLENIUS Z. VARGA M. KRACHLER L. FONGARO J. ZSIGRAI *Joint Research Centre* European Commission
554	Application of data center technology in the digitalization of nuclear security systems	QING LIU *Shenzhen Newcross Company Limited* China
584	Virtual reality system for simulating the nuclear facility security system in Indonesia	JEHAN AL RAMADHAN RIZAL ALFARIJI URSULA KRISNA UTAMININGTYAS SETYO ASIH *University of Gadjah Mada* Indonesia
597	Identifying, developing and promoting best practices in nuclear forensic science: The Nuclear Forensics International Technical Working Group (ITWG)	M. CURRY *United States Department of State* United States of America M. WALLENIUS *Joint Research Centre* European Commission
617	Enhancing security through integrity verification of operating systems using blockchain on physical protection system	SEUNGJUN LEE HYUN YANG GUHEUM YEON *OPCIA Corp.* Republic of Korea
631	Building security culture in newcomers: A model training program based on risk concern	F. INCE B.G. GÖKTEPE *NUTEK Inc.* Türkiye
644	Diversity equity inclusion and accessibility: A national laboratory perspective	V.N. VARGAS D.B. HARMON *Sandia National Laboratories* United States of America
657	Open framework for the evaluation of the level of maturity in physical and cyber security capacity for facilities with nuclear or other types of radioactive material	G. PUERTA Colombia J.M. VILANUEVA J.P.P. LOZANO *Ministry of Mines and Energy* Colombia J.S. ROJAS Colombia

Abstract ID	Title of presentation	Contributor(s) and affiliation(s)
671	Management of a radiological incident involving an orphan source discovered in a container returned to port	M. TCHAOU *University of Kara* Togo E. HALIBA E. HAZOU C.Y. FIAGAN *University of Lomé* Togo
674	Societal culture influence in nuclear security culture: Understanding societal cultural dynamics to proactively mitigate security risks	P. LYNCH *Lawrence Livermore National Laboratory* United States of America C. SPERANZA *Oak Ridge National Laboratory* United States of America Y.V. IBRAHIM *Centre for Energy Research and Training* Nigeria P. BIANCHI *Nuclear and Energy Research Institute* Brazil L. GABBAY *Israel Atomic Energy Commission* Israel E. GOSSET *French Alternative Energies and Atomic Energy Commission* France J. WALKER *Indiana Tech University* United States of America
688	Smart packaging for critical energy shipments	S. HOLLIFIELD MINGYAN LI M. MATHUR K. MONDAL G. PHAM I. SIKKEMA *Oak Ridge National Laboratory* United States of America
692	3S regulatory inspections — Radioactive sources	A.A. HAMMAMI *Federal Authority for Nuclear Regulation* United Arab Emirates

Abstract ID	Title of presentation	Contributor(s) and affiliation(s)
711	Strengthening the security regime through technical cooperation on sustainable development in Latin America and the Caribbean	F. DELUCHI *Argentina Global Foundation* Argentina J. FERRER *National Atomic Energy Commission* Argentina

Programme Committee:

A. Solano Ortiz [1] (Chairperson)	Costa Rica
C. Danestig Sjögren [1] (Chairperson)	Sweden
V. García Gutiérrez [2,3] (Chairperson)	Costa Rica
A. Nilsson [2,3] (Chairperson)	Sweden
M. Roston [1]	Argentina
M. Via [1]	Argentina
P. Zunino [3]	Argentina
K. Kirakosyan [3]	Armenia
D. Manukyan [1]	Armenia
A. Mikayelyan [3]	Armenia
M. Francis [1,2,3]	Australia
J. Lee [1,2,3]	Australia
M. Vieira [1]	Brazil
M. Da Silva Giafferi [1]	Brazil
C. Romão [1,2,3]	Brazil
A. Roza de Lima [1,2]	Brazil
Y. Dimitrova [2,3]	Bulgaria
P. Endezoumou [3]	Cameroon
A. Simo [3]	Cameroon
A. Breton [1]	Canada
C. Cochrane [2,3]	Canada
L. Hartery [2,3]	Canada
N. Semblat [1,2,3]	Canada
Chen Chen [2,3]	China
Jingmin Ju [3]	China
ZiPing Li [1]	China
Hong Lu [2]	China
Qingming Wei [1]	China
C. Guillermet-Fernández [1]	Costa Rica
L. Villalta [1]	Costa Rica
O. Abdelgawad [1,3]	Egypt
M. Hamdy [1]	Egypt
N. Nawwar [1,2,3]	Egypt
J. Galy [1,2,3]	European Union
M. Nauduzaite [2,3]	European Union
T. Hack [1]	Finland
F. Cottel [1]	France
V. Lesage [1]	France
S. Paultre [2,3]	France
W. Rother [3]	Germany
H. Kroeger [2]	Germany
S. Adu [1]	Ghana

R. Agalga [2,3]   Ghana
S. Dampare [3]   Ghana
S. Karri [3]   India
S.B.P. Ayyappan [1,2]   India
F. Bacconnier [3]   International Criminal Police Organization (INTERPOL)
M. Jugheli [3]   INTERPOL
G. Sedda [1,2,3]   Italy
J. Frischknecht [3]   Japan
Kazuko Hamada [2]   Japan
Naoko Kuribayashi [1,2,3]   Japan
Z. Molnar [1,2]   Japan
Shigaeki Sato [3]   Japan
Joung Hoon Lee [2,3]   Korea, Republic of
T. Zayerh [1,2,3]   Morocco
S. Jonah [1]   Nigeria
O. Okiti [2,3]   Nigeria
U. Ghiyas [1,3]   Pakistan
A. Shakoor [1,2,3]   Pakistan
A. Vigil Puga [1,2,3]   Panama
I. Olea [1]   Philippines
D. Umingli [1]   Philippines
A. Dedu [1,3]   Romania
M. Georgescu [3]   Romania
E. Blagodarina [1,2]   Russian Federation
D. Bokov [1,2,3]   Russian Federation
A. Bulychev [1]   Russian Federation
S. Marogulov [3]   Russian Federation
V. Ostropikov [1,3]   Russian Federation
A. Althomali [1]   Saudi Arabia
M. Ashary [1,2,3]   Saudi Arabia
M. Majozi [1]   South Africa
A. de las Casas Fuentes [1,2]   Spain
F. Yllera Sanchez [2,3]   Spain
M.H.M. Abuuznien [1]   Sudan
A. Franzén [1,3]   Sweden
B. Laggner [1]   Switzerland
M. Oberle [1,3]   Switzerland
R. Straub [3]   Switzerland
B. Ozyardimci [2,3]   Türkiye
E. Soylu [3]   Türkiye
M. De Carlini [1]   United Kingdom
C. Claydon [3]   United Kingdom
B. Daly [2,3]   United Kingdom
T. Eagleton [1]   United Kingdom
O. Shofoluwe [2,3]   United Kingdom
P. Smith [2,3]   United Kingdom

N. Sobey [1]	United Kingdom
C Sommerstein [2, 3]	United Kingdom
F. Miorin [2]	United Nations Interregional Crime and Justice Research Institute
M. Lorenzo Sobrado [1, 2, 3]	United Nations Office on Drugs and Crime
S. Ayers [2]	United States of America
M. Dingerdissen [1]	United States of America
L. Duchene [3]	United States of America
A. Fragoyannis [1, 2, 3]	United States of America
A. Hallock [1, 2, 3]	United States of America
S. Minot Asrar [1]	United States of America
S. Neakrase [3]	United States of America
P. O'Brien [2]	United States of America
Hongliu Zhang [1]	1540 Committee

Note:

1 — Attended the first Programme Committee meeting, held in Vienna on 1–3 March 2023.

2 — Attended the second Programme Committee meeting, held in Vienna on 21–23 November 2023.

3 — Attended the third Programme Committee meeting, held in Vienna on 27–29 February 2024.

IAEA Secretariat:

Scientific Secretary:	S. Mroz
Scientific and administrative support:	B. Denehy
	C. Deura
	S. Atayeva
	R. Fikare
	S. Frolov
	S. Henry Bolt
	Sen Li
	Lai Peng
	P. Rosol-Barrass
	H. Sears
	V. Tafili
	D. Tutunjyan
Event organizers:	S. Padmanabhan
	E. Paniagua-Miranda
	L. Popescu
	M. Neuhold

CONTACT IAEA PUBLISHING

Feedback on IAEA publications may be given via the on-line form available at:
www.iaea.org/publications/feedback

This form may also be used to report safety issues or environmental queries concerning IAEA publications.

Alternatively, contact IAEA Publishing:

Publishing Section
International Atomic Energy Agency
Vienna International Centre, PO Box 100, 1400 Vienna, Austria
Telephone: +43 1 2600 22529 or 22530
Email: sales.publications@iaea.org
www.iaea.org/publications

Priced and unpriced IAEA publications may be ordered directly from the IAEA.

ORDERING LOCALLY

Priced IAEA publications may be purchased from regional distributors and from major local booksellers.

25-04272E

Printed and bound by CPI Group (UK) Ltd, Croydon, CR0 4YY

06/07/2026

02160600-0009